TESTIMONIALS

"Nicole Strickland is a well-respected member of the paranormal community who tirelessly researches claims of alleged ghosts and hauntings. Her passion for the field is what separates her from the rest of the pack. The plethora of knowledge she possesses regarding the history and hauntings of the RMS Queen Mary is not only astounding but also quite admirable. Nicole is the 'go-to' source for information about The Grey Ghost (nickname used for the Queen Mary *during World War II)."*

—Joey Tito, author and paranormal researcher
with the American Spectral Society

"Nicole is my 'go-to gal' for Queen Mary *history and ghostly activity. Her approach is both professional and informative. Having worked with Nicole several times in the past, she has never disappointed nor displayed anything but pure heart, experience, and professionalism. I highly recommend reading her work and booking her for your next radio interview or conference! One of the best in the field and truly one of the community's sweethearts!"*

—Todd Bates, radio host and EVP specialist

"Nicole Strickland is not merely a ghost hunter. This is not a person that you'll find searching for random bumps in the night, but a true researcher of the unknown, the paranormal, and the unexplained. Not content to sit back and simply read about the work done by others, Nicole is out actively investigating the mysteries that still exist when it comes to ghosts and hauntings. There are few who can match her passion and drive for answers about the supernatural and I consider her one of the most dedicated researchers in the field."

—Troy Taylor, author and founder of American Ghost Society

"Nicole brings her serene energy to investigations, and it is absolutely a pleasure to work with her. She has a reverence and respect for the spirits. She is respectful and considerate, not only to the spirits, but also to the people she is with. Nicole has captured some phenomenal EVP, many of which have been caught while on board the Queen Mary, *her favorite haunt. She has educated herself on the history and happenings on the Queen of the Seas, and we have enjoyed several evenings with her on this wondrous ship, as well as investigations in Southern California. We've shared experiences on a personal level, regarding the death of a cherished family member. Just recently, Nicole had a significant experience and when she told us about it, imagine our amazement in that it completely coincided with a personal experience that we both (Dawn and Sharon) had shared together. It was enlightening and comforting. Nicole is humble about her intuitive experiences, and we look forward to further investigations and shared encounters in the future with her."*

—Dawn Gaudette and Sharon Gaudette Hieserich,
intuitives and paranormal researchers

"As History Ambassador for the San Diego History Center, I am always on the look-out for experts in their field who can provide another means to learn about the past. Paranormal investigation, and delving into Spiritualism, is always a hot button item for the public, especially as San Diego claims to possess America's most haunted house. This claim is part of the folklore of the region, and storytelling is the most compelling way we can engage with myths, legends, and tall tales—the essence of folklore.

Nicole Strickland has been the paranormal researcher and storyteller of choice for my guests over the years as she understands the genre deeply and shares key clues, identifies strategies, and demonstrates equipment used in the investigation process that is engaging, thought-provoking, and an intelligent way to approach what might otherwise be a skeptical subject for some people. She puts the science of paranormal investigation in the hands of the participants so they are actively immersed in the experience to whatever degree they feel comfortable.

Nicole is a passionate paranormal researcher and author, the founder of the San Diego Paranormal Research Society, and is sought out nationally to serve on panels, radio shows, and television interviews in regards to this topic. As she is so thorough in her approach to every investigation or request for understanding of the paranormal phenomenon, Nicole imparts assurance and instills confidence in her clients and participants in your audiences.

She also shows empathy to the energies and spirit forces that she might come across in her experiences. She realizes the gift that she has been given to serve as a connection between two worlds and doesn't take that honor lightly. As a result, she will tell you directly that she is not a ghost hunter. Though she doesn't deny that there are many avenues to pursue when connecting with the spirit world, her approach is one expressing understanding, a willingness to listen, and to offer assurance to those who are unable to fully cross-over. These are the characteristics that make Nicole highly sought after and what makes her such a vital resource in the paranormal community today."

—Gabe Selak, History Ambassador, San Diego History Center

"If it came down simply to her successful writing career and foundational work in the field of ghost research, I would find plenty of reasons to admire my friend Nicole Strickland. She is an exceptional person whose drive, laser-focused energy and gift of gliding prose justify her booming popularity.

But there is another trait Nicole possesses which sets her apart from others in the crowded field of paranormal investigation, and that is her passionate affinity for places and people whose stories echo history across time and oceans.

Just absorb her works on the legendary RMS Queen Mary and you will come as close to time travel as you are likely to experience. The Haunted Queen of the Seas; Spirited Queen Mary: Her Haunted Legend and RMS Queen Mary: Voices From Her Voyages all create a smartly readable look at seafaring opulence during peacetime and an urgent repurposing of a world-class luxury ship during the Second World War.

Meticulously researched and studded with narratives that read like love letters to a legend in repose, Strickland's works allow the departed souls of the ship's crew, guests, and military personnel take readers page by page on a brisk cruise through the various eras of the Queen Mary *as "a prodigy of the seas," to borrow the author's phrase for a vessel still shimmering today in retirement at California's Long Beach Harbor. The big boat endures as elegantly as ever but apparently is stacked from hull to upper decks with an eerie historical residue told in captured whispers by a host of restive voices. In their mirth and mourning, anger and angst, these highly distinct spirits noisily persist in a safe harbor, united in devotion to a glorious past.*

—Gary Mantz, Co-host, *Mantz and Mitchell*, 1150 KKNW Seattle

Photo courtesy of Joe Bertoldo

Haunted Queen of the Seas

The Living Legend of the RMS Queen Mary

By Nicole Strickland

Published by KayliMax Books
Fifth Edition, November 2024
ISBN: 979-8-9880654-3-2

Published by KayliMax Books

Interior layout by Rachel Newhouse for elfinpen designs
Cover design by Amelia Greene for Penoaks Publishing

www.authornicolestrickland.com

Maria Regina mari me commisi
(*Queen Mary* committed me to the sea)

DEDICATION

This book is dedicated to all of my family, friends, and colleagues who have supported and encouraged me throughout my life. I love you all so much.

I also dedicate this book to all those who believed in the future of the RMS *Queen Mary* and to those who helped her have a successful career; to all of her former crew and passengers; and the many World War II servicemen and servicewomen who fought for their lives to help bring peace to the nations.

To Dillon, Ralph, and Will. You are a part of the *Queen Mary's* legacy.

To the RMS *Titanic* and all her souls who touched upon her decks.

To John, little Jackie, and the many *Queen Mary* spiritual residents, as you have taught me so much. As always, I wish you peace and love.

ACKNOWLEDGMENTS

I want to thank all of those who have helped this book become a reality. Writing this book has been a dream waiting to happen.

I want to thank my San Diego Paranormal Research Society friends and colleagues for their support and encouragement.

Special thanks to all those who have contributed to this book.

A very special thanks to *Queen Mary* enthusiast, Joe Bertoldo, for his many gorgeous photographs of the ship.

Thank you to all the *Queen Mary* spiritual residents who have eagerly shared their stories. Thank you for contributing to the *Mary's* living legacy.

CONTENTS

FOREWORD

When my mind wanders to thinking about the RMS *Queen Mary*, I not only think about her magnificence; I also immediately think about Nicole Strickland. I have never encountered someone so magically intertwined with the entire essence of the *Queen Mary* more than Nicole.

With a somewhat otherworldly energy, she intrinsically knows about the many, many layers of the ship. Whether traveling through the thought-provoking and fascinating time periods of the liner's history or exploring a paranormal encounter, Nicole can extrapolate the information with an unparalleled ease.

I find it noteworthy to say that this book will allow you to embark on the *Queen Mary's* many voyages with such detail that you will no doubt find yourself swept up in the images and words as if you were traveling aboard firsthand.

Nicole Strickland is a master journey teller, with her innate ability to share with the reader an inexplicable supernatural occurrence in such a way as to create a real-time energy usually reserved for the experience itself.

Having personally experienced paranormal incidents aboard the *Queen Mary*, I find it most refreshing that Nicole goes to great lengths to accurately and practically describe encounters without any sensationalism or grandiose descriptors. Her passion and adoration for the *Queen Mary* elicits the deepest and utmost respect for those who have sailed on the legendary ship and continue to walk the halls from the past.

I know for a fact this book will leave you with a feeling of excitement, curiosity, fascination, and wonderment about the *Queen Mary*. You have in your hands annals of historical and paranormal happenings from eras gone by that will leave you wanting to visit the historic vessel for yourself. Do not be surprised if upon a visit you encounter a former passenger or crew member who is more than happy to knock on a wall as you pass by. Or perhaps a visit to B Deck will be greeted with a whispered "hello" in your ear, its speaker unseen with the naked eye.

Are you ready to step onto the decks of the *Queen Mary's* unprecedented history and unparalleled paranormal occurrences? Come aboard, turn the pages, and let Nicole take you on a remarkable odyssey aboard the RMS *Queen Mary*.

Cher Garman
The Chicago Files

NUMBER 534

For ages you were rock, far below light,
Crushed without shape, earth's unregarded bone.
Then Man in all the marvel of his might
Quarried you out and burned you from stone.

Then, being pured to essence, you were nought
But weight and hardness, body without nerve;
Then Man in all the marvel of his thought,
Smithied you into form of leap and curve;

And took you, so, and bent you to his vast,
Intense great world of passionate design,
Curve after changing curving, braced and mast
To stand all tumult that can tumble brine,

And left you, this, a rampart of a ship,
Long as a street and lofty as a tower,
Ready to glide in thunder from the slip
And shear the sea with majesty of power.

I long to see you leaping to the urge
Of the great engines, rolling as you go,
Parting the seas in sunder in a surge,
Shredding a trackway like a mile of snow

With all the wester streaming from your hull
And all gear twanging shrilly as you race,
And effortless above your stern a gull
Leaning upon the blast and keeping place.

May shipwreck and collision, fog and fire,
Rock, shoal and other evils of the sea,

Be kept from you; and may the heart's desire
Of those who speed your launching come to be.

—John Masefield, "Queen of the Sea"

QUEEN OF THE SEA

Hearts will glow, with admiration,
When our new liner leaves the quay.
And the name, loved by the nation,
Will give her charm and dignity.
British labour gave its skill...
and it's giving me a thrill ... cause
I've booked my trip for the USA
So when I go over the sea
The Queen Mary *takes me.*

I'm happy and gay,
cause I'm sailing away
I've booked my trip for the USA
On the finest ship in the world
The Queen Mary ... *how'd you like to come with me?*
The ship is all British, it's wonderful too,
The ship is manned by a British crew,
So when I go over the sea,
The Queen Mary *takes me.*

There'll be fun galore
and people I adore
That's why I'm happy and gay,
cause I'm sailing away
On the finest ship in the world
The Queen Mary ... *Queen of the Sea*

—By Horatio Nicholls, a.k.a. Lawrence Wright
(Maiden Voyage, 1936)

"There's no doubt about it; ships do have a soul. It is not a question of sentiment. When I see this old girl going to the states, well, it is the next best thing. It would be pure sacrilege for this great, lovely ship to go to the breakers. It just couldn't be done."

—Tom McCarthy Hamilton, RMS *Queen Mary* Librarian

"I, as Captain of this ship, am very proud and honored to take her on her last voyage. It is with some sadness that she is sailing for her last time, but I am glad that I am not taking her to the scrap yards and that in Long Beach she will be a living monument to British shipbuilding, engineering and seamanship."

—Capt. John Treasure Jones, RMS *Queen Mary* Captain

"We'll be sad to walk away and leave such a beautiful ship in a foreign land, but it is fitting that this ship should go to America. Americans are the only ones who can keep her grandeur intact."

—William Trueth, RMS *Queen Mary* Steward

"To many of the passengers the Queen Mary *had become just that: my ship."*

—William J. Duncan, author of
RMS Queen Mary: Queen of the Queens

"The Queen Mary, *pride of Great Britain, had endeared herself to millions of American soldiers, and, indeed, the entire free world owed her a great debt of gratitude."*

—*The Official Pictorial History: The Queen Mary*

INTRODUCTION

A Reigning Love for the Queen of the Seas

"The best and most beautiful things in the world cannot be seen or even touched—they must be felt with the heart."

—Helen Keller

Photo courtesy of Joe Bertoldo

In 2005, when I first drove down the stretch of road called Queens Highway, I was awed by my real-life view of the stateliest ship of all time, the RMS *Queen Mary*. As I pulled into the massive parking structure adjacent to the *Mary*, I was not yet aware of how much I would come to truly love this ship.

When I got out of my car, I was standing and facing toward the massive ship, frozen by her beauty and character. I was scanning my head left and right as I proceeded toward the *Mary*. Once I touched upon her decks, I could feel her rich history and abounding character seeping from her walls. I describe the feeling as similar to being drawn into a time slip and being one among the many jubilant passengers who once sailed on the ship. I checked into the onboard Hotel Queen Mary and was transformed back in time as I was walking down the long hallway toward my stateroom on A Deck.

After I settled in my stateroom, A024, I decided to walk around the *Mary's* various decks that were unrestricted to the public. As I ascended the stairs onto her Sun Deck, I was mesmerized by the panoramic views of Long Beach. I remember standing on her starboard side and thinking of all the many individuals who have traveled aboard the Queen during her days on the seas. I believe it was at this very moment that I came to love this ship and connect to the *Mary's* soul.

I have traveled to the *Mary* many times since my very first visit back in 2005. Each time I touch upon her decks, I develop an even closer kinship to the Queen of the Seas. I personally feel so connected to this grandeur of a ship and I feel at home when I am aboard. I am sharing the fact that I feel so connected to the *Queen Mary*, as if the liner is a part of me.

I have often pondered why I feel the love that I do toward the *Mary*. I have often wondered if others feel the same way toward her as I do. Well, I have learned through my various readings about the ship's history that many of her former passengers and crew have come to love her just as much as I do. There is just something mystical about the *Queen Mary* and it has been this way since her inception. She embraces you and connects with people in a very special way.

Many people may feel that the love of the ship is due to her unrivaled history on the seas. However, I feel there is more to it than that. There is more to it than we can comprehend. There is abundant life and character to the *Mary*, which has been felt since the dawning of her days. Even though her reigning days on the seas are over, her soul still speaks. She is very much still living.

I traveled to the ship in July of 2009 with some friends when one of my intuitively-inclined buddies told me that she needed to talk to me regarding the *Queen Mary*. As she and I were taking a stroll on the ship's Sun Deck, she told me something magnificent about the *Mary*. She told me that she felt as though the ship is alive, complete with emotion and thought. I marveled at my companion's prophetic words as they echoed exactly what I felt the very first time my feet crossed the threshold onto the *Mary's* A Deck. It was at this point that I started to fully realize that other people can feel the life of the *Mary's* soul as well.

My friends and I were only able to stay aboard the *Mary* for one night. The following morning as we were slowly walking toward our car, I started to get the sadness that always pervades me when I have to say goodbye to the ship. As my eyes started welling up with tears, I turned to my friend P.J. and said, "I am always sad to leave this ship, and I cry every time that I do." P.J. looked at me and said, "She knows you'll come again." Her words could not have been any more accurate as I visit the majestic liner at least once per month.

I must add that there is a strong spiritual aura surrounding the RMS *Queen Mary*. It goes beyond the fact that she was a ship designed to bring peace to the nations. There were so many near-death incidents and tragedies that the *Mary* miraculously overcame during her reigning days on the seas. The vessel emanates so much love and strength, which undoubtedly helped build a protective bubble around her. Even though she is forever docked in Long Beach, she still projects a powerful sense of love, strength, and peace. To this day, the *Mary's* soul still whispers to us. If you listen closely and open up your heart, you can hear her.

As a professional paranormal investigator and researcher, I truly believe that places become paranormally active or haunted for various reasons. I firmly believe that one cannot separate a location's history from its ghostly events. History and the paranormal are closely connected. They are relatives with a dear kinship. In terms of the RMS *Queen Mary*, her abounding history is still alive and still whispers to us. In the pages to come, I will share some of her misfortunes and successes from her unrivaled 31 years of history on the waters.

There are various explanations as to why the RMS *Queen Mary* is noted as one of the most haunted places in the world. Then there are the reasons we cannot decipher as to why she is so haunted, which further add to her mysticism. Former psychic and *Queen Mary* paranormal investigator Peter James claimed that the ship holds hundreds of spirits. On my visits and overnight stays aboard the *Mary*, I have encountered some incredible paranormal events, which will be elaborated in the pages to come. Many of these encounters were of a spiritual nature as if the ship was speaking to me and openly sharing her unmatched history on the oceans.

As I relayed above, the *Mary* has endured many misfortunes and successes while on the seas. It must be noted that the *Mary* survived various obstacles, which could have been her end of days. However, like a Phoenix rising out of fire, she indeed survived every single one of them and continued to triumph.

The *Queen Mary* also made it through World War II and was instrumental in shortening the war and lengthening the lives of many young men, women, and children. The *Mary's* successes have also made her go down in history as being one of the most legendary vessels to ever cross the seas. Her history is so rich and full, clearly making it impossible to cover in just one book. However, I will do my best to give a colorful and detailed overview of the *Mary's* history and paranormal events. This book is written from the heart and soul, and in writing this manuscript I want to help keep her living legend alive.

The RMS *Queen Mary* has always been known as a legendary ship. She is still alive and well even though her sailing days are over. British psychic Mabel Fortescue-Harrison predicted that the *Mary* would become exactly what she is today: a very famous ship. On the day of the *Mary's*

September launching in 1934, Fortescue-Harrison said, "Most of this generation will be gone, including myself, when this event occurs; however, the *Queen Mary*, launched today, will know its greatest fame and popularity when she never sails another mile and never carries another passenger."

This is my tribute to the RMS *Queen Mary*. So sit back and relax as you read about the world's most famous and legendary ship of all time. This book is my way of thanking the *Mary* for all that she has done in her lifetime and all that she continues to do; it is a quest to help highlight her magnificent living legend.

"TRANSCENDENCE"

Our RMS Queen Mary knows well the oceans of the world, but her travels move beyond the boundaries of bodies of water on earth, as she lives, breathes, and sails on through the world's heart, mind, and spirit.

By Norma Strickland

PART 1

The Dawning of the Queen of the Seas

"There is no courage in entering upon great enterprises in prosperous times, but I have faith in the future, and confidence that the Cunard Company will hold its own upon the Atlantic."

—John Burns, Cunard Company Chairman

Photo courtesy of Joe Bertoldo

The dawning of the RMS *Queen Mary* started with the strong vision of and belief in the future of steamships. Young merchant Samuel Cunard first envisioned the idea to build a vessel back in 1831, when he watched a steamboat approach a port in Halifax, Nova Scotia. Cunard was a business partner with his father, and he pondered if a steamboat could travel the Atlantic Ocean equipped with both people and supplies. In 1833, Samuel Cunard bought the same steamship he saw enter the port of Halifax back on that summer day in 1831. *The Royal William* made its way into Gravesend, England, carrying both cargo and passengers nearly a month later. This signaled the birth of Cunard's steamship company.

Young Cunard believed that adequately constructed steamships on the North Atlantic could be just as effective as the use of railroad trains. In 1838, Cunard persuaded the British government into believing that steam was more reliable than sails and was therefore granted the first steamship contract. The new Royal Mail Steam Packet Co. would offer two monthly voyages for mail service between Liverpool, Halifax, and Boston.

In fact, Cunard may not have even been able to develop the British and North American Royal Steam Packet Company if he had not come into contact with an 1838 British government advertisement looking for bids to develop a twice-monthly steamship postal service between England, Halifax, and the United States.

Three steamships were eventually constructed, with the first being the *Britannia*. The *Britannia* was 207 feet long and weighed 2,000 tons. Her engine ran at a speed of just over 700 horsepower. She also utilized her sails to change her power and conserve her fuel. This small liner set the example for future Cunard steamships. Some of the ships that followed in the *Britannia's* wake included *Carmania, Mauretania, Lusitania,* and *Aquitania* as well as the *Queen Elizabeth* and *Queen Mary*. A second *Mauretania* and the *Caronia* also followed.

The *Britannia* left Liverpool in the summer of 1840 and carried both passengers and mail between England, Boston, and Halifax, Nova Scotia. The ship made the voyage twice every four weeks under the Admiralty

Contract. She was only 230 feet long as compared to the extreme size of the *Queen Mary*. The *Britannia* and her sister ships were all constructed in Clydebank, Scotland, which definitely established an unprecedented tradition of shipbuilding.

In 1881, the *Servia* joined the Cunard fleet of ships. She was the first to be constructed of steel and include Thomas A. Edison's recent electric lighting invention. She was an express liner that carried a large passenger capacity and traveled at a speed of approximately 17 knots.

John Brown and Company was probably the most established shipyard in the entire world. Its origins began in the Finnieston area of Glasgow in 1847. Two brothers, James and George Thomson, originated a marine engine-building business at the Clyde Bank Foundry. The brothers were both mentored by engineer expert, Robert Napier, who helped to bring marine engineering and iron shipbuilding to the Clyde.

In 1851, James and George Thomson founded the renowned shipbuilding company, which covered five acres. In 1852, the brothers developed their first ship. The Clyde Bank Iron Shipyard earned a strong reputation for constructing quality passenger vessels, many for the Cunard line. Tragically, in 1863, an ensuing argument caused James to leave the business. George continued on but sadly passed away in 1866.

A trust was set in place to run the company until George's sons were old enough to take over. In 1871, the corporation relocated to the Barns o' Clyde (Clydebank), which made way for the expansive launching of ships. In 1899, John Brown and Company, a Sheffield steel-maker, took over the shipyard, having built many famous liners prior to the construction of the RMS *Queen Mary*.

Government-subsidized foreign shipping companies were becoming popular among the trans-Atlantic trade. On May 28, 1930, Cunard declared that John Brown Shipyard in Clydebank, Scotland, would be responsible for the construction of the 1,000-foot ship that was initially labeled as Job 534. Engineers built a distinct 500-foot tank and started to experiment with models of ships. Engineers spent hours simulating every single aspect of North Atlantic weather. More than 8,000 prototype experiments took place prior to the selection of the 17-foot, 800-pound, self-propelled model, a miniature version of the actual ship.

In the early 1930s, the RMS *Queen Mary* surely represented an elusive and brave responsibility. The ship was to be the biggest, fastest, and most luxurious liner ever designed. In November of 1930, the small model went through its last trial runs. The Cunard Steamship Company was impressed by the model and accepted its design. Thus, John Brown Shipyard was given a $30-million contract on December 1, 1930, for the building of Job 534. Job 534 represented an unprecedented challenge of shipbuilding and expressed due faith in British ship design and engineering expertise.

Construction of Job 534 commenced in November of 1931 with the diligent work of the liner proceeding ahead of schedule. With 7,000 workers meticulously constructing the *Mary's* inner walls and hull, Cunard declared that the she would be ready for launch into the sea in May of 1932. Construction of the vessel took place during the day and at night in Clydebank. However, on December 11, 1931, the work of Job 534 was put to a halt and over 3,000 workers became unemployed. The Depression directly affected Cunard and the building of its ships. Travel to and from America to Europe was restricted due to the declining economy.

During this time, Sir Percy Bates was now the Cunard board chairman and insisted that the work of Job 534 continue. He even assured England that the ship would indeed be finished. Percy received mail on a daily basis, which contained small contributions from those who desperately wanted to see Job 534 in its final state.

Great Britain's Neville Chamberlain begged for a government subsidy and loan to continue construction of the liner. The skeleton of the ship sat in a long-idled two-and-a-half year wait. The British government eventually stepped up to the plate and subsidized the finishing of Job 534. Ultimately, Cunard was coerced to merge with the White Star Line as a condition of government assistance. Thus, it is no surprise that many, many people were eternally proud of the vessel and looked forward to the future development and supremacy of British-constructed ships on the North Atlantic.

The building of Job 534 resumed schedule on April 3, 1934. Accumulated rust was scraped off the hull and thousands of nesting rooks were removed from cranes and scaffolding. Workers and village people sang "The Cunarder's Restarting" to the tune of "The Campbells are Coming." Workers happily marched toward the shipyard as townspeople cheered and joyfully shed a tear or two. In other words, the Scots were eternally grateful for the re-continuation of the prized liner. The *Mary*, in its completion, weighed 81,237 tons and had 12 decks, which reached the height of the Statue of Liberty's hand. At the time, she was the largest ship in the entire world.

H.M. Tomlinson wrote a preliminary article to be placed in the Cunard-White Star Company pamphlet, which celebrated the birth of the RMS *Queen Mary*. Tomlinson witnessed the ship as she was being built and he wrote down his impressions in that very same article. The following are some excerpts from his article:

> *"The Queen Mary was a complete surprise. There could be no argument with that superior and gracious presence. Majesty was there, for her lines were beautiful. Size is but size, but great magnitude is forgotten in nobility and form.*
>
> *"In this immense ship there shows a joy and faith in mechanical power which have transfigured it to beauty. Her very prow, looking down at an observer with the haughty indifference of the sphinx—the hawse openings are very like eyes, downcast—is superior to questioning.*
>
> *"The Queen Mary is a personality with her own heritage and attributes.*
>
> *"She was born of the science and skill of so many men concerting for a single purpose, and out of tradition and experience reaching so far into the past, that only a skilled examiner may read the signs of her heritage.*
>
> *"She is the result of a world of creative influences. She belongs to the ways of peace."*

Llewellyn Roberts and Lewis Roberts were the *Queen Mary's* chief engineers when she was nearing her completion. Once the hull of the liner was set in place, work was started on the *Mary's* interior appointments. After sailing down the Clyde in the springtime, the ship remained in dry-dock for 18 days. Approximately 80 engineers along with electricians, plumbers, boilermen, and refrigerating engineers assisted in prepping the grand vessel's completion. In fact, 4,000 men were employed to work on the ship. In the last few months of the *Queen Mary's* construction, many staterooms and cabins were built. Other rooms, including those reserved for officers and engineers, were all marked out.

On March 24, 1936, Job 534 was ready to leave the shipyard, and a million people came to see her off to the sea. Timber creaked as the vessel acquired speed. Fifty-five seconds later, 200,000 people witnessed the *Mary* come to rest in the river 1,196 feet away. Tugboats helped assist the giant liner as it was nudged into the narrow channel of the River Clyde. The *Mary* proceeded down the Clyde in two tides. The first tide came early in the morning. Due to the ship's tremendous size, the Clyde was widened and expanded at the mouth of the River Cart, which was a branch of the Clyde itself.

When the ship reached the mid-channel, her massive propellers began to move the ship downstream. The shipyard and both banks of the river were covered with cheering crowds who came to wish "God Speed" to their newest and most luxurious creation. Children even had a day off of school to watch this historical event. Family and acquaintances were even invited to come witness this super liner as it entered her element.

However, her launch into the water wasn't without potential tragedy. This incident foreshadows many more potential near-death experiences the ship would endure. She ran aground two times, but fortunately did not sustain any damage. It was said that if the ship had been stuck on land in her position for 20 minutes or so, the water level would have fallen, which would have prevented her from coming off the land. Worse, she could have been forever fractured and a total loss. However, she prevailed and continued on with her sea trials. Furthermore, the liner was expected to be insured for the approximate sum of $35,000,000.

Under the auspices of John Brown engineers, the ship continued with her sea trials for the next few weeks and was grateful for the decent weather conditions. Tugboats and tenders surrounded the ship each morning to bring last minute provisions and passengers onboard before pulling up her anchor and traveling up and down the Firth of Clyde for her trials. One of the crew members relayed that the scenery and ambience during the ship's sea trials were memorable. The liner would pass by many small islands in the Firth of Clyde before she made her way to Southampton to undergo final preparation for her scheduled maiden voyage, which would take place on May 27, 1936.

More than 7,000 tests were completed with model ships in small tank oceans before the final design was selected from a committee of shipping men. The committee decided on using propulsion for the new vessel. The committee of men consisted of Sir Aubrey Brocklebank, Sir Charles Parsons, Vice Admiral R.W. Skelton, Sir Thomas Bell, Andrew Laing, T. McPherson, G. McL. Patterson, Andrew Hamilton, J. Callender, Commander C.W. Craven, and John Austin.

The experiments included diesels, diesel electrics, direct turbine, turbo-electric, and geared turbines. The committee selected the single-reduction geared turbines, whereby power supply would come from 24 water boilers of the Yarrow type. The water boilers created high pressure and high temperature steam. It was surmised that the turbines would create 200,000 horsepower. Furthermore, the *Mary* was designed so she could refuel in an eight-hour period.

Basically, the *Queen Mary* ran by a system consisting of geared turbine engines, which were supplied by steam from its high-pressure boilers. Twenty-seven huge boilers supplied steam and covered a space of five rooms. The ship's four propellers were driven by four turbines. Each of the 257,000 turbine blades was tested and integrated by hand. There were four sets of engines. The propellers were individually casted from fifty tons of manganese bronze. The four gear wheels were 14 feet each in diameter. All together, they weighed approximately 320 tons. It is a fact that the main machinery area aboard the *Mary* is so huge that the area by itself is bigger than most ocean vessels of her time period.

On her last sea trial run, the *Queen Mary* traveled from Southampton, her first trip under the Cunard-White Star flag. Six hundred renowned guests traveled aboard for the 24-hour cruise of a little less than 500 miles. Sir Percy Bates and Sir Edgar Britten both welcomed the guests.

King Edward VIII and his royal party visited the liner while it was docked in Southampton. The royal party that accompanied the King were the Duke and Duchess of York, Princess Elizabeth, the Duke and Duchess of Kent, the Duchess of Gloucester, and, of course, H.M. Queen Mary. Commander Sir Edgar Britten even discussed the ship's various design features with the King.

A grand ceremony was thrown in honor of the ship's launching on September 26, 1934. At this point, the ship was still known as Job 534 and no one knew what her new name would be, except for the King himself. He meticulously kept the liner's name a secret until it was finally revealed during the ceremony. H.M. Queen Mary cut the satin cord and crashed a bottle of Australian wine against the ship's bow as she said the following words: "I am happy to name this ship *Queen Mary.*"

The crowd was surprised and joyfully cheered. H.M. Queen Mary went on to say, "I wish success to her and to all who sail in her." The *Mary* gracefully moved toward the river. As the luxurious vessel approached the sea, hundreds of small boats, including paddle steamers and row boats, traveled along beside her. Airplanes were flying overhead, carrying eager reporters and photographers. The Royal Marine Corps band played "Rule *Britannia*" as the grand ship approached the ocean.

According to Sir Ashley Sparks, resident director of the lines in the United States, not one person in the Cunard-White Star Company knew of the Queen's decision to name vessel 534. No one knew about her decision until she stepped up to the microphones and disclosed the vessel's new name to the world.

"The surprise was an exceptionally pleasant one, for it is impossible to imagine a more fitting name," said Sir Ashley Sparks.

In fact, the Queen broke two ship naming traditions when she named liner 534. Cunard liners had names that always ended in "IA" and White Star liners had names ending in "IC." Many thought that the Queen would name vessel 534 after the *Britannia* for symbolic purposes.

People strongly envied those who traveled on the *Queen Mary* for her maiden voyage. The morning of the ship's first transatlantic voyage was clear and called for good weather. Auras of excitement permeated throughout the ship. Friends and relatives were there to say goodbye to loved ones. Stewards were running around like crazy adhering to the Cunard tradition of excellent service. The ship's voice blasted in the air as people on the ground cheered the liner on her way into the Atlantic. Indeed, it was a day to never be forgotten.

Thousands of people came to Southampton the night before the *Mary's* maiden voyage, which was scheduled on May 27, 1936. H.M. Queen Mary came to inspect the vessel the day before its maiden departure. Five trains ran from Waterloo Station carrying the ship's 2,000 passengers. More trains held London spectators going to get a glimpse of the grandest ship ever designed as she made her way toward New York. The *Mary's* maiden voyage sailing time was 4:30 p.m. The Royal Marines Band played music as the liner sailed into the sea. Radio kept track of her hourly progress.

The *Mary's* giant whistles blasted and echoed throughout the air as they signaled the beginning of the liner's first passenger voyage. According to credible sources, the ship averaged a speed of approximately 28.5 knots on her maiden voyage from Southampton to New York. Therefore, she did not attempt to claim the coveted Blue Riband prize. However, the ship was said to most likely travel at a faster speed on her second voyage.

Nearly 2,000 passengers were onboard for the ship's maiden voyage. The crew consisted of 1,186 men and women. Additionally, there were more than 100 reporters and journalists. Twenty broadcasters from five countries performed live broadcasts from various locations throughout the ship. The crew loaded thousands of mail sacks en route to the United

States and Canada. Many distinguished passengers were aboard for the RMS *Queen Mary's* maiden voyage.

A film crew was onboard to capture the memorable moments of the liner's premiere voyage. The film, *Wonder Ship,* was developed, released, and displayed in numerous theatres throughout the country. Additionally, one of the most popular orchestras of the time was onboard to add to the overall splendor of the *Mary*. Each night, dance music could be heard via wireless communication.

On the second day of the *Mary's* maiden voyage, a dense fog closed in around the liner. Commodore Britten was immediately struck by the memory of the *Titanic's* maiden voyage disaster in 1912. He slowed the *Queen Mary*, which ultimately prevented it from beating the *Normandie* and taking the title of the fastest ship on the seas. As the stately Queen sailed by the exact spot of the *Titanic's* demise, she slowed down just enough to allow crewmen to toss over a flowery wreath in memory of those who lost their lives on that tragic morning in 1912.

After four days at sea, the *Queen Mary* made its way toward New York City. She had been at sea for four days, five hours, and 46 minutes. The liner was escorted to Pier 90 by a circus of Navy boats, including Coast Guard, motor, sail, and speed boats. The *Mary* lowered her anchor when she arrived at the Quarantine Station. She had to await customs, health, and immigration officials. Once she was cleared, she made her way into the Port of New York and went on to Pier 90.

The Cunard line representatives expressed their gratitude in the way the New York Harbor Police Department handled the massive crowd of people who came to see the *Mary* as she entered the port. Pier 90 is the second of the group of five 1,100-foot piers, which the Dock Department envisioned several years prior The *Mary* was to be berthed along the north side of Pier 90. The building of these piers started in 1931 with the first of the pier groups opening to the *Normandie.*

As for the *Queen Mary*, Pier 90 had a 400-foot wide and 46-foot deep basin, which provided berthing space for the liner. Additionally, it was equipped to handle freight service and accommodate passengers. Five passenger elevators with a 4,500 capacity were installed for passenger convenience. People could board and disembark a ship without the frustration of steep gangways. Engineers have called Pier 90's terminal the greatest marine terminal in the world.

New York was getting ready to welcome the *Queen Mary* as photos of the ship appeared in various restaurant and store windows. The police unit distributed 481 uniformed individuals as well as placing detectives in the vicinity of Pier 90 and the ship's berth. Longshoremen worked diligently as they cleaned and decorated the pier.

The New York harbor was full of excited spectators. While the ship was stationed in New York on her maiden voyage, she was open to the public. Thousands of guests touched upon the *Mary's* decks for private parties. Ten thousand people even received a tour of the ship. The liner was opened to the general public for the very first time at 9:00 a.m. From opening time until noon, 2,570 people toured part of the *Mary*. Additionally, 3,288 more people visited part of the ship between the hours of 2:00 p.m. and 4:00 p.m. that same afternoon. Approximately 5,858 people paid $1.00 to board the *Mary*.

The Blue Riband is a prestigious prize given to a passenger ship with the fastest east- or west-bound North Atlantic crossing. The *Queen Mary* attained this accolade twice in her career on the seas. In August of 1936, the *Queen Mary* captured the Blue Riband from the *Normandie* on her sixth voyage. According to Sir Percy Bates, the quick speed the Queen liner attained to receive the coveted prize was absolutely necessary in order to see how much faster she could sail beyond the 28.5 knots.

The liner crossed the Atlantic in three days, 23 hours, and 57 minutes from Ambrose Lightship to Bishop Rock on the western edge of England. The ship had an incredible speed during this approximate 3,000-mile voyage at 30.63 knots. What an incredible rate for such a large ship. Every other ship in the harbor saluted the *Mary* as she confidently sailed into Southampton. She had made the fastest round trip ever completed across the North Atlantic. Sir Edgar Britten felt that the *Mary* would hold the victory in her hands for many more years.

"We have still got a few more horses left in our horsepower bag if we want them," explained Chief Engineer Llewellyn Roberts.

The Southampton Harbor Board spent $100,000 on the dredging of the liner's ship operation site. Instead of utilizing the new ports in Southampton's harbor, the *Mary* used the ocean dock. On her way to Southampton, thousands of onlookers and various small boats celebrated as they were notified of the ship's new record by radio personnel. Radio station WOR and the Mutual Network broadcasted when the liner passed the lightship.

It wasn't until 1937 that the *Normandie* re-captured the prize from her predecessor. After some time in dry dock to receive adjustments and brand-new propellers, the *Queen Mary* won back the coveted Blue Riband in August of 1938 on her 48th round-trip voyage. She finished the 2,907-mile westward journey in three days, 21 hours, and 48 minutes, at an impressive rate of 30.99 knots. The *Mary* held onto this accolade for the next fourteen years, until the S.S. *United States* achieved it in 1952.

The *Mauretania* captured the record in 1910 while traveling at 26.06 knots and was able to hold onto it until July of 1929. In 1929, the *Bremen* made the voyage from Cherbourg to the Ambrose Channel lightship in just over four days. The *Bremen* held onto the prize until August of 1933. The *Rex* was next in making top speed and held onto the accolade until June of 1935.

It was not just the fine craftsmanship and engineering skills that enabled the Cunard ships to be the finest vessels in the world. According to Cunard, the people who help run the ships are equally important to their overall luxury and success. The men and women who helped run the *Queen Mary* had some of the highest credentials and were carefully selected to uphold Cunard's time-honored tradition of excellent service and accommodation.

White Star fleet's navigating officers held a master's certificate and were formerly qualified to take command of the ship. Twenty-four of the captains were officers in the Royal Naval Reserve. Twenty-two captains had the Royal Decoration and six of them were Officers of the Order of the British Empire. Furthermore, two captains had the Distinguished Service Cross. One of them was even a Baronet.

The *Queen Mary* held approximately 110 officers. Eighty-four of the ship's officers were even engineers. The officers' function and rank were distinguished by the bands of gold lace on the lower part of the uniform's sleeve. The different colors of the slim velvet strip in the middle of the gold lace bands also differentiated the officers.

Sir Edgar Britten was the RMS *Queen Mary's* first commodore and had an extensive list of responsibilities. Britten was ultimately accountable for the entire ship and every life contained within it. He was also responsible for keeping a strict time schedule.

Other types of crew members were just as important to the overall success of the *Mary*. The *Queen Mary's* Engineering Department contained a chief engineer and a staff chief engineer. Under the administration of the chief and second electricians were the watch-keepers or third electricians. There were different ranks of the ship's electricians. Most of the crew members responsible for catering to the passengers were the stewards, stewardesses, waiters, cooks, and chefs, which totaled over 1,000 people.

Even though there were low paid wages in the British Merchant Service, the RMS *Queen Mary* was the highest paying liner. However, the money earned still wasn't at its prime. Crew members were paid according to their level of service with the stewards being the most prosperous.

Millions of people revered the work and craftsmanship of the RMS *Queen Mary*. The ship was the epitome of strength, beauty, and stability. Sir Percy Bates expressed that the *Mary* was a vessel of peace. He also elaborated by saying that the *Mary* would be the most lucrative liner to ever set sail.

The *Queen Mary* marked her 2nd year at sea while she sailed on her way to Plymouth, Cherbourg, and Southampton. On May 26, 1938, she carried 1,500 passengers and plans were made for a midnight celebration in honor of the ship's anniversary. Officials declared that the *Queen Mary* had a total passenger list of 116,787 persons during those first two years at sea. In fact, the astounding statistic was a world record for that particular time period. The liner's total passenger count for her two years at sea was distributed according to class. She carried 37,267 cabin class passengers, 41,223 tourist class passengers, and 38,298 third class passengers. The vessel completed 86 crossings and set a mark of 2,716 people per round trip.

On September 27, 1938, the *Queen Mary's* sister ship, the *Queen Elizabeth,* was launched at John Brown's shipyard. The *Queen Elizabeth* ended up being ten feet longer than her sister ship. She also had 14 decks, as opposed to the 12 decks on the *Mary.*

Even though the *Mary* had unrelenting success during her career, her travels were not without unfortunate passenger injuries. In October of 1936, thirteen people were physically hurt on the ship as it traveled from Southampton via Cherbourg during a severe storm. Passengers received injuries due to falls while the ship encountered a sixty-mile gale and a heavy beam sea. Staff Captain B.H. Davies relayed that the gale lasted for four hours. Thankfully, the liner was not destroyed by the powerful storm.

Another unfortunate event befell the stately ship a few days later. On October 28, 1936, Sir Edgar Britten was found dying on his cabin room floor. Apparently, Britten suffered a stroke. He was carried ashore and taken to a nursing home. His wife rushed to be with him; however, he died before she reached his bedside. Britten was 62 years old at the time of his tragic demise. He was due to retire the following April.

With flags at half mast, the *Mary* entered the Cherbourg harbor, saddened by the sudden death of the well-admired Britten. Sir Arthur Rostron, a retired commodore of the Cunard-White Star Line, commemorated the lost captain by saying that Britten was one of "the most sensible and the most reliable men I have ever come across." Captain Reginald V. Peel was then mandated to rush to Southampton to take command of the *Mary.*

Sir Edgar Britten was born in Yorkshire, England, in April of 1874. He attended a few years in grammar school. He later went to sea in the bark *Jessie Osborn.* Britten continued to work on ships until he acquired his master's certificate. He joined the Cunard line in 1901 as Fourth Officer of the *Ivernia.* At a fast pace, Britten ascended the ranks and became master of various ships. The vessels he commanded were *Campania, Franconia, Laconia, Andania, Ascania, Mauretania, Aquitania,* and *Berengaria.* King George V knighted Britten in February of 1934.

Greyer times lay ahead for the RMS *Queen Mary* as World War II was imminent. The *Mary* contained a record number of passengers as she sailed on her last commercial voyage on August 30, 1939. A profound aura of extreme anxiety was quite pervasive for the peacetime travelers on this last luxury voyage prior to World War II.

The massive ship eventually arrived in New York harbor after war was declared and was berthed there for the winter time. The *Queen Mary* and *Normandie* were now positioned as neighbors in New York harbor, silently awaiting their next task at sea. The *Mary* acquired an excessive amount of silt as she lay in port for six months. Thus, sand-suckers worked on getting rid of the silt that was nestled around the liner's keel.

In March of 1940, the British Government informed Cunard that the *Queen Mary* would be requisitioned for service in World War II. People would not see the traditional Cunard colors of red, black, and white for six years. Every square inch of the *Mary* was painted grey as an attempt to keep her camouflaged from the enemy during the war.

On March 7, 1940, the *Mary's* running mate, the *Queen Elizabeth,* came into New York's harbor awaiting her wartime duties. The *Queen Elizabeth* was the newest and biggest ship, while the *Normandie* was the most luxurious. The *Queen Mary* was the fastest liner. What these three ships had in common was the long silent wait before their wartime duties.

The month of March was devoted to preparing the *Queen Mary* for World War II service. In addition to the generous slap of grey paint, the letters spelling out her name were removed. Cunard-White Star ordered that all of her luxury items be removed from the ship in New York in order to prepare for troop conversion. Over 200 cases of crystal, china, and silverware were secured, packed, and put into storage. For the length of the war, the *Mary's* expensive and luxurious items were secured in Cunard warehouses in New York.

Seven-hundred-and-seventy British merchant marine officers and seamen arrived via the Cunard-White Star liner *Antonia.* Many of the men were transferred to the *Mauretania* and the rest were taken to the *Mary*. No information regarding these men was disclosed.

The *Mary's* colossal size made her a constant target, so meticulous precaution was ordered to make sure the stately ship remained afloat. American intelligence reports relayed that Nazi agents in the United States premeditated on blowing the liner to pieces or setting her on fire. Thus, she was equipped with security appointments and her pier was also heavily patrolled. In fact, the *Queen Mary*, along with the *Mauretania* and 24 other British ships, was recognized as an "enemy warship" if the German Navy ran into her while at sea.

The *Queen Mary* received the installation of an ASDIC underwater sound detection system. Her fan-tail also received a single four-inch gun that would help protect her from smaller ships. Lewis and Vickers machine guns that were utilized during World War I were secured onboard the *Mary*. The liner was also outfitted with a degaussing girdle, which deactivated magnetic mines.

In the early morning hours of March 19, 1940, *Queen Mary's* Captain R.B. Irving received notification that the ship had to be ready to set sail within 48 hours to reach Australia. Thus, the ship was equipped with 470 officers and men from the *Antonia*. Her lifeboats were being tested and 500 officers and seamen from the *Antonia* were ordered to begin duty on the *Queen Mary*.

There was much speculation as to the exact route the *Mary* would take on her way to Australia. The *Mary* was too wide to effectively pass through the Panama Canal. Thus, it was more likely that the ship would sail on an eastern route, which would take the liner around Cape Horn and through the Pacific.

On the morning of March 21, 1940, the *Queen Mary* once again set sail downstream toward the sea. She was escorted by the Coast Guard and the New York Police Department. Once she reached the vast Atlantic Ocean, she moved southward until she became invisible.

For the next six years, the luxurious RMS *Queen Mary* retained her role as a warship. In her camouflaged grey color, she would cross many oceans carrying troops, war brides, and children to and from their destinations. During the period of World War II, the *Mary* would endure many difficult times and would see many untimely deaths. Without a doubt, the war period would turn out to be the most complicated and trying time of her entire career.

However, before delving into the RMS *Queen Mary's* wartime role, I would like to talk about the types of luxury that she offered to her many passengers. In many ways, she was considered a floating palace, offering something to everyone who boarded her. The following section will discuss the *Mary's* various luxurious appointments, amenities, and passenger accommodations.

A Luxurious and Stately Queen of the Seas

"We make a living by what we get, but we make a life by what we give."

—Sir Winston Churchill

Photo courtesy of Joe Bertoldo

ART DECO AND DESIGN

The RMS *Queen Mary* was conceived to be the fastest and grandest ship to ever sail the North Atlantic. Thus, the art appointments aboard the *Queen Mary* were some of the finest of their time and were suggestive of the Art Deco era. In fact, there is so much art aboard the *Mary* that many people overlook some of the minor detailing that added to the ship's stately ambience.

At the time of her building, the *Mary* was the epitome of revolutionary marine design and expertise. As Cunard affectionately referred to the ship as "Britain's Masterpiece," a committee was established to oversee the liner's entire design process. Sir Percy Bates chose American architect Mr. B. V. Morris as his individual consultant on the decorative plans for the ship. While sojourning at Mr. Bates' England residence, Mr. Morris perused the vessel's initial plans. Mr. Arthur Davis, a London-based architect, was also called upon to assist in the collaborative effort in designing the ship.

Once the room layouts were chosen and accepted, the committee was tasked with recruiting various artists and craftsmen. All in all, a team comprising some thirty artists, sculptors, painters, and interior designers was called upon for the creative mission. The *Mary's* art was decorative and displayed mythology, flowers, animals, gardens, and landscapes. These various works from noteworthy artists were placed throughout the ship.

One of these notable artists was Doris Zinkeisen, who was responsible for designing the Verandah Grill, a room aboard the ship that embodied sophistication. Zinkeisen was an artistic celebrity in her own right, appearing in countless columns and magazines as well as creating costumes for film productions. For the Verandah Grill, Zinkeisen chose a carnival-like theme that represented a mixture of the theater, circus, night club, and ballet. In my opinion, it is one of the most ingenious and imaginative Art Deco murals ever conceived.

Another popular feature exists in the Grand Salon, which served as the main dining hall for cabin class passengers. Macdonald Gill's huge decorative map of the Atlantic showcases the mobile crystal models of the *Queen Mary* and *Queen Elizabeth*, specifically depicting their exact position on the sea at any given moment. It has been said that one post-war captain refused to dine in the room if the models were not working properly, thus suggesting how integral each art piece was to the *Mary's* multi-layered tapestry.

In the 1920s and 1930s, artists and interior designers had various artistic styles and technologies at their fingertips. Germany was responsible for starting the trend of luxurious passenger ships. Charles Mewes was a French architect and master designer who had much creative influence in artistically outfitting these stately ships. Many of the new luxury liners in the 1920s and 1930s provided service and accommodation similar to grand international hotels. Mewes' design expertise profoundly changed the design and decoration of ships and revolutionized passengers' experience.

Design proposals for the suites and cabins were more complex as several models were constructed. It was hard to make the final selection of the scheme that would best fit the ship's overall Art Deco magnificence. The *Mary's* interior compartments were adorned with a combination of furnishings, such as wood paneling, carpeting, and artwork, which established an admirable and tranquil feeling. Throughout her walls, she was coated with the finest finishes of rare woods taken from all of His Majesty's colonies. Her bathrooms were covered with formic, which was a new and affluent material in 1936. The lighting fixtures were intricately placed to accentuate the ship's overall luxurious feeling. Furthermore, miles of soft flannel were laid in the ship to prevent any creaking sounds when the ship was at sail. The finest selections of British decorative arts adorned her 27 public rooms.

The grandest art selections were displayed in the extravagant first class public rooms: the Dining Room, Main Lounge, Drawing Room,

Ballroom, Smoking Room, Observation Bar, and Verandah Grill. In addition to the fine and rare woods, beautifully constructed textiles and fabrics embellished artistic accents and animation. The *Mary's* artistic design definitely accentuated the fact that she was a luxurious and stately ship.

The *Mary* utilized 56 of the entire world's grandest and rarest types of woods. The liner's first class main entrance was paneled in Masur birch, which was cross-banded with plain birch and trimmed in English Elm Burr Dado. The first class restaurant was adorned with light and dark Peroba with Maple Burr panels. The *Mary's* main staircase was covered with White English Mottled ash accompanied with English Elm Burr Dado. Chestnut-oak and plain oak walls could be seen in the first class shopping center located on Promenade Deck. For the various rooms and passageways, attractive ornamental woods were utilized. Some of the other types of exceptional woods that were used consisted of Makore, Acero, Angelim, Avodirbe, Black Bean, Bubinga, Corbaril, Ebony, Maidu, Padouk, Synara, Thuya, and Zebrano. The following is a list of some of the types of woods used:

Acero: A maple type of wood found in Europe, which was used in the cabin staterooms.

Angelim: A type of wood found in South America, which was used in the alleyway on Main Deck.

Ash: This type of wood hails from the British Isles and Europe. It was used in the cabin staterooms and tourist restaurant. Olive ash comes from England and was used in the cabin staterooms and tourist library.

Avodire: A type of wood generating from West Africa, which was used in cabin staterooms.

Beech: An English wood displayed in cabin staterooms.

Betula: A wood of Canadian birch used in the cabin staterooms.

Masur Birch: Stems from Eastern Europe and was used in ship's paneling. It was the most difficult wood to use.

Black Bean: A type of wood that is native to Australia and used in cabin staterooms.

Bublinga: A type of wood coming from West Africa and was used in the Observation Lounge.

Cedar: This wood comes from Honduras and covered the ship's cigar shop.

Cedarmah: A fusion of Cedar and Mahogany, which was used in the Observation Lounge.

Cherry: Seen in the Main Deck corridors and stems from West Africa.

Chestnut: This type of wood comes from the Far East and can be found in the ship's main locations.

Corbaril: A rare type of wood that can be found in many areas of the ship.

Ebony: This attractive wood comes from Macassar and Celebes Islands. It was used in the Observation Lounge and some particular suites.

Elm: English elm was used in many areas of the ship.

Greywood: This wood hails from India and was used throughout the ship.

Laurel: This wood stems from India, Burma, and Ceylon and was used in the Starboard Gallery.

Mahogany: This type of wood has many varieties and comes from Honduras, India, and Nigeria. It was used in the majority of staterooms.

Maidu: It was used on cabin furniture.

Makore: This wood stems from West Africa's Gold Coast and was used on cabin furniture.

Maple: Maple comes from all over the world and was used in various parts of the ship. Different types of maple were used in particular areas of the *Mary*.

Myrtle: Stems from the U.S. Pacific Coast and used in the cabin staterooms.

Oak: Comes from British Isles and found throughout the ship.

Padouk: This wood comes from Burma and Adaman Islands and can found in the tourist Smoking Room.

Pear: An English wood, which was found in the cabin writing rooms.

Peroba: Stems from Brazil and used in the cabin restaurant.

Pine: Hails from various parts of the world and used in staterooms.

Rosewood: Stems from South America and Bombay and found in cabin suites.

Satinwood: A Ceylon and East Indian wood found in the panels in the cabin staterooms.

Sycamore: An England-based wood used throughout the ship.

Synara: A French type of wood used in the cabin children's play area.

Teak: A Burma type of wood used throughout the ship.

Thuya: A North-African wood used throughout the ship.

Walnut: Used on furniture and in cabin staterooms.

Yew-Tree: Used for furniture and in special suites.

Zebrano: Stems from the West Coast of Africa and used in certain suites.

The ship's main dining area was bigger than the grand ballroom in the Waldorf-Astoria, an opulent hotel located in New York City. The dining facility had reserved space for up to 800 people and displayed a rare type of wood on its ceiling and walls called Peroba. Peroba stems from Brazil and is a particularly modern type of wood, which had never been utilized in a ship nor was it used on land at the time. The dining room's metal components were constructed out of bronze with the fiber being covered with Ruboleum.

The staterooms themselves were equipped with appointments that added to the *Mary's* luxurious setting. Tall mirrors were positioned in a way that would move to a person's every angle. Lights were installed in long vertical wardrobe closets. The stateroom's clock was timed from the liner's clock headquarters. The various carpet, curtains, and chair coverings were complete with interesting fabrics.

As mentioned above, the works of many famous artists were featured in the ship. One of the most notable works is the large ornamental map of the North Atlantic, which was positioned at the forward end of the Grand Salon. This map displays both the Old World and the New World with the RMS *Queen Mary* uniting them. The four private dining facilities had their own distinct paintings, which gave them a characterized perspective. The most popular painting was a circus scene by Dame Laura Knight. The Children's Room contained various murals depicting nursery life. The tourist and third class rooms also displayed grand pieces of art. Below is a list of some of the artists and their works that contributed to the *Queen Mary's* decorative design:

Lady Hilton Young (Lady Kennett)—Marble plaque of H.M. Queen Mary, which is located at the head of the staircase facing the main hall.

Mr. Cedric Morris—Two flower-study paintings at both ends of the starboard gallery.

Mr. George Ramon—Incised and painted designs on the Children's Playroom walls.

Mr. Macdonald Gill—Decorative map of the North Atlantic in the Grand Salon on R Deck.

Mr. Bainbridge Copnall—Applied carvings on wood, which are located in the dining area.

Mr. James Woodford—Three pierced and carved screens in the Smoking Room as well as the modeled symbolical figures in the Smoking Room.

Mr. Norman F. Forrest—Four seasonal statuettes.

Additionally, Sir Percy Bates' dear friend, Rudyard Kipling, consistently encouraged him throughout the course of building the *Queen Mary*. Thus, Kipling was called upon to come up with a Latin dictum for the *Queen Mary* medal. Once he devised it, Mr. Bates asked the finest

Latin intellectuals to review the text. The chosen phrase reads *"Maria Regina mari me commisi,"* translating to "*Queen Mary* committed me to the sea."

DINING OPTIONS

The RMS *Queen Mary* offered some of the finest varieties of cuisine in the entire world. In fact, Cunard trained its chefs in the world's top restaurants prior to employing them as chefs aboard its luxury liners. These culinary gurus were trained to master the making of various types of cuisines. Each class of passengers enjoyed sumptuous varieties of food. Luncheon and dinner menus differed from day to day and offered about 80 dishes.

Many private dining facilities were available. However, the majority of tourists preferred the Main Dining Room where they could be the center of attention. Many of the world's wealthiest stars dined in the Main Dining Room. Some of these stars included Fred Astaire, Greta Garbo, Bob Hope, Elizabeth Taylor, Clark Gable, and Gloria Swanson. The Captain's table contained the most prominent seats in the *Queen Mary's* Dining Salon. The captain himself selected seven favorable individuals at the start of a voyage to dine with him at his table.

The Verandah Grill accommodated 82 first class passengers, and it was encouraged that travelers make reservations weeks in advance. The Verandah Grill contained its own private bar and kitchen. At midnight, the grill became a luxurious night club called the Starlight Roof where passengers could indulge in dancing and entertainment.

Breakfast included 80 options of food, from the standard ham and eggs to eleven different cereals to five types of toast, among others. The first class evening meal was a seven-course dinner, which was served in the *Queen Mary's* Dining Salon. Today, it's known as the Grand Salon on R Deck. It was the largest room ever constructed within a ship.

The 1936 breakfast menu—shown in this book's photo spread—contained options such as fruits, cereals, fish, eggs, omelettes, entrées, potatoes, salads, cakes, breads, and beverages. The menu even offered an assortment of cold meats. Passengers could select entrée options, such as minced chicken creole or roast beef in cream. Some of the grill options included tomato and Oxford sausage, York and Wiltshire ham, mushrooms on toast, ham steaks and chutney sauce, and calf's liver and bacon. The beverage portion offered tea, coffee, Cadbury's cup chocolate, Horlick's malted milk, cocoa, Ovaltine, instant Postum, or Kaffee Hag coffee.

Passengers could even select from an extensive assortment of alcoholic beverages. The ship offered 31 proprietary brands of whiskey. Seven different gins were available. People could choose among the 14 types of cognacs. Beers, wines, and a large amount of champagnes were offered as well. The ship also carried teas from Ceylon, India, and China.

The following are some statistics pertinent to the cuisine of the *Mary*: 77,000 pounds of fresh meat, 27,500 pounds of chicken, 11,000 pounds of fresh fish, 66,000 pounds of potatoes, 33,000 pounds of fresh vegetables, 15,000 dozen eggs, 22,000 pounds of flour, 11,000 pounds of sugar, 1,300 gallons of milk, 1,110 boxes of various fresh fruits, and 4,400 quarts of ice cream. When looking at the *Queen Mary's* past menus, it is obvious that cuisine options have changed over time in tune to what peoples' palates desire.

Today, the *Queen Mary* has kept her tradition of excellent cuisine as she offers award-winning restaurants, including the prized Sir Winston's, which is known for its beef wellington. Sir Winston's Restaurant offers the most upscale dining options and a cozy ambience. Not presently in service, The Promenade Café had various dining selections and probably the best fish and chips that I have ever had, and I have had my share. Chelsea Chowder House & Bar offers delectable seafood cuisine as well as many varieties of American fare. It cooks up the best patty melt that I have ever tasted! The Observation Bar features a generous beer, wine, and cocktails menu, including an array of mouth-watering appetizers.

PASSENGER AMENITIES AND ACCOMMODATIONS

People chose to travel on the *Queen Mary* for many obvious reasons. She was a ship of luxury and extravagance and believed to be the speediest liner of her day. Passengers received royal treatment the minute they touched upon the ship's decks. Baggage personnel hospitably took passengers' belongings to their staterooms. Bellboys courteously escorted the passengers to their cabins. Travelers took great pride in sailing on this floating city of elegance.

The *Mary* attracted different types of people. Many Americans traveled aboard her decks. Hollywood film stars and producers, bankers, industrialists, and other business moguls sailed on the grand liner. Members of the aristocracy, diplomats, and wealthy Englishmen traveled on the *Mary*. The ship also carried individuals of nobility and royalty. Those that had the opportunity to cross the Atlantic on the liner definitely had a reason to brag and boast. Indeed, the *Queen Mary* helped to forge relationships between people of various cultures.

The types of passenger accommodation and amenities were plentiful for each class, to say the least. Thus, the cabin, tourist, and third classes (the class labels after World War II became first, cabin, and tourist, respectively) all benefited from the *Mary's* opulence and lavishness. The tourist and third class passengers received esteemed services that far exceeded those that were offered on other liners. There were various services and facilities offered to each class, which included barber and beauty salons, dining rooms, gymnasiums, bars, smoking rooms, cinemas, and shopping facilities.

As already mentioned, the *Queen Mary's* interior rooms were decorated with a plethora of furniture, wood paneling, carpeting, and artwork, which set the foundation for a desired ambience and mood. Critics and journalists praised the ship's interior décor. In fact, the *Mary* was known as the "Ship of Beautiful Woods." There were so many amenities available to each group that one could think of the *Mary* as a town or city.

The main lounge offered many different activities for cabin class passengers. Organ recitals and relaxation took place in the mornings and afternoons were reserved for tea. After dinner, people enjoyed dancing until midnight, accompanied by the orchestra. Catholic Mass was held in the cinema on Sundays and in the drawing room on other days. The *Queen Mary* was the first liner to offer Jewish worship in a small synagogue located aboard.

Exercise facilities were abundant on the *Queen Mary*. One of the most talked about amenities was the R Deck first- and third-class swimming pool, which was available to passengers from 7:00 a.m. to 2:00 p.m. and 5:00 p.m. to 7:00 p.m. The pool eventually opened up to third class passengers when air travel became the primary choice for the affluent in the 1960s. People could enter the swimming pool via the revolving door onto the balcony or from the Turkish bath area. This pool room contained golden quartzite on its ceiling, which was utilized by the Egyptians. The ship actually had two swimming pools that were both indoors. In bad weather conditions, the ship's rocking could affect the pools, so their usage was obviously discouraged during severe weather. Sun Deck even had a gymnasium, which was open for twelve hours.

Hundreds of clocks were displayed throughout the grand liner. About four-hundred clocks were placed in the first class staterooms and seventy clocks were placed in the public areas. B Deck enclosed areas and rooms devoted to passenger accommodation as well. According to *The Queen Mary: Her Early Years Recalled,* some of the cabin (first) class luxury accommodations included the following:

> **Gentlemen:** Shaving, haircutting, shampooing, dry shampooing singeing, oil shampoo, oil shampoo and vibro, scalp massage,

tonic dressings, mudpacks, high frequency massages, lilac and peroxide application, and hand massage

Ladies: Permanent waving, finger waving, marcel waving, waving and curling, trimming, trim and singe, ordinary shampoo, tar shampoo, henna shampoo, chamomile shampoo, spirit shampoo, hand and head massage, vibro head massage, hot oil massage, high frequency massage, bleaching, steam scalp treatment, brushing, and dressing

Beauty Treatments: Hand and face massage, vibro face massage, high frequency massage, eyebrow arching, mudpack Boncilla and Egyptian, Magnesia pack, peroxide and olive oil massage, manicuring, hand massage and finger stretching, frictions (Coty and Houbigant), frictions (Atkinson), and Chiropody

The Turkish and Curative baths were located on C Deck. Different types of rooms offered various types of therapy and even included relaxation cubicles. Included were a frigidarium, tepidarium, massage room and electric bath room, steamroom, calidarium, laconicium, and attendant's room. According to *The Queen Mary: Her Early Years Recalled,* some of the therapies in this category included the following:

Electric bath with Alcohol Rub Infra-red Irradiation
Turkish bath with Alcohol Rub Ultra-violet and Infra-red
General Massage
Diathermy
Local Massage
X-ray Photograph and Exam
Ultra-violet Irradiation

The *Queen Mary's* Promenade Deck is located directly underneath Sun Deck. Chairs were dispersed throughout the deck so passengers could relax and enjoy the scenic ocean views. Additionally, people could walk around the Promenade Deck, which was equivalent to walking approximately 1/3 of a mile. Furthermore, this area contained many of the ship's public rooms, such as the Observation Lounge and Cocktail Bar.

Entertainment was offered to passengers every night. Even the officers were permitted to attend the nightly festivities. As the evening hours approached, the Queen Mary Concert Orchestra would conduct an orchestral musical show in the Main Lounge. During dinner, passengers enjoyed the works of the Queen Mary Dance Orchestra as well. At 7:30 p.m., people could enjoy cocktail music. The Starlight Roof Club also featured late-night dancing. Occasionally, pre-general release movies were shown as well as some of the great classic films. People even had the opportunity to enjoy a game of Bingo in the Main Lounge.

The ship housed other important public rooms, designed to suit the interests of artists and entertainers who traveled on the ship, which were located on the portside of the liner. A studio was available for artists to practice their entertainment pieces in a sound-proof room. A lecture room was even provided for artists to show and discuss their works. Travelers could utilize the Writing Room, which contained a vast range of books. The library also housed more than 1,400 standard and modern English titles, and 250 of them came in nine languages. It also contained quite a few modern fictional works and universal literature pieces.

The liner also contained public rooms located on its starboard side. A Drawing Room, Writing Room, and Children's Playroom were frequently used. In fact, this playroom was one of the most intriguing locations ever to be conceived on a luxury liner. It included a boys' and girls' section, an aquarium, and a model cinema theatre. Its ceiling had artwork displaying the sun, a man-in-the-moon, and many shining stars. The liner also had a children's nursery as well.

The *Mary* also incorporated other essential services to accommodate the satisfaction of the passengers. A branch of the Midland Bank was available so tourists could take care of their financial and monetary matters. Travelers even had the option to discuss their nutritional diet preferences with the restaurant manager or head waiter. People were encouraged to order their dining cuisine choices well in advance. Deck chairs, cushions, and rugs were accessible and could be acquired through the deck steward. Travelers on the *Queen Mary* usually came equipped with expensive attire. Thus, it makes sense that the ship offered a clothing pressing room. Canines were even permitted onboard, but not allowed in the various public rooms or on passenger decks. In fact, 80 feet of deck space was provided so the dogs could exercise on Sun Deck. Furthermore, a mail office was available so passengers could arrange for their mail and postal packages. A safe deposit box was also installed on A Deck to help preserve tourists' belongings and valuables, which were often of the expensive kind.

Directly from the ship, tourists had the option to speak to Great Britain, France, Germany, Spain, Holland, Belgium, Sweden, Switzerland, Denmark, Austria, Canada, and the United States. Multi-lingual stewards were utilized as a passenger convenience. The lapel on their coats indicated their country and native language. Additionally, at 6:15 p.m., a British news broadcast could be heard in the Long Gallery. The American news broadcasted at 6:30 p.m.

The cabin class consisted of very wealthy individuals. In total, the *Queen Mary* had 26 first class suites. The cabin class suite was obviously the most expensive stateroom on the ship, equipped with beautiful furnishings, decorations, and bragging rights. These suites usually contained four rooms and came in various sizes. The sizes varied from one master bedroom with two smaller rooms all the way to two master bedrooms with four smaller rooms. The centerpiece of each suite included a sitting area and a parlor with comfortable couches and chairs. Each centerpiece had paintings, handmade carpets, and ceramic sculptures. An electric fireplace was also present. A maid's room was also included with each suite as well.

Accommodation for the cabin class passengers was provided in staterooms, which were located on Sun Deck, Main Deck, A Deck, and B Deck. Each of these cabins had two single beds. They also contained a private bath equipped with a tub, shower, and sink. The tub had the elements of hot and cold salt water and fresh water. In the 1930s, salt water was believed to be quite therapeutic. Furthermore, the majority of the cabins resembled an L-shaped pattern to effectively utilize the width of the ship. These staterooms displayed a modern style to appeal to passenger comfort. The cabin class staterooms on A Deck and B Deck contained furniture that was made out of various woods with veneer enhancement. Writing tables and portable folding tables were provided in the majority of the cabins.

The tourist class bathed themselves in luxury and extravagance as well. They also had an indoor swimming pool, gymnasium, and spacious deck areas for recreation and exercise purposes. Second class passengers could relax in two lounges and dine in an air-conditioned salon. Additionally, this class was offered a smoking room, library, and writing room. Children traveling in the tourist class also enjoyed playing in the second class Children's Playroom. This group of tourists had the option of browsing items in two shops. A bank was available to them as well as a barber shop, bureau, and other service departments.

The tourist class staterooms spread out over five decks and depicted luxury just like the cabin class staterooms. The cabins could fit up to four people comfortably. These passengers were given the option of having hot and cold water. Many rooms contained individual showers and toilets. The rooms' furnishings included mahogany bedsteads with lee-board and tea trays. These staterooms offered generous closet space. Second class stateroom accommodations surpassed that of other luxury liners.

The third class travelers were also offered conveniences and accommodations that far exceeded those of other liners. They could also choose from a variety of daily activities. An assortment of cuisine choices were made available as well. Their cabins were extremely nice, comfortable, and cozy.

For the three years following her maiden voyage, the *Queen Mary* carried passengers from all walks of life on the North Atlantic run. Soon, however, the joyful glee and carefree lifestyle aboard the ship was to be overshadowed by the imminent dark clouds of war. The stately liner made her last peacetime sailing in August of 1939. The next few years ultimately tested the *Mary's* strength, endurance, and courage, as she was tasked with the perilous responsibility of being a troop ship. For the entire duration of this unforgettable and horrific battle, the *Queen Mary* continued to prove her reign on the seas. It was during this time that she and her sister ship, the *Queen Elizabeth,* helped save the lives of numerous men, women, and children while also forging a healthy relationship between the United States and Britain.

It is here that I will reiterate Winston Churchill's notable words regarding the Cunard sisters' efforts during World War II: "Built for the arts of peace and to link the Old World with the new, the Queens challenged the fury of Hitlerism in the battle of the Atlantic. Without their aid, the day of final victory must have unquestionably been postponed."

PART 3

The RMS Queen Mary's Fighting Role

"One ought never to turn one's back on a threatened danger and try to run away from it. If you do that, you will double the danger. But if you meet it promptly and without flinching, you will reduce the danger by half. Never run away from anything. Never!"

—*Sir Winston Churchill*

Photo courtesy of Joe Bertoldo

Commodore Gordon Illingworth's October 1943 message to the troops:

> *"I call upon all officers and men to obey my orders to the letter. I have but one task. It is the job of bringing this ship safely to port, and that job, God willing, I will do. It is not important that you, numbering some 15,000, arrive safely in the Firth of Clyde, but it is important that the ship be brought safely to anchor there. Remember that. You and I are not indispensable to the successful prosecution of this war, but the ship is. You will keep in mind, therefore, that all your thoughts during the crossing will be directed toward her security. Enemy forces will be at work, and the Hun will try every device in his power to bring the 'Queen' to harm. Submarines will trail us and aircraft will harass us. They have done it before and we have every reason to believe they will do it again. But the 'Queen' will take care of herself. From now until the moment you debark, think in terms of the ship. Treat her gently and do not abuse her. She stands ready to do for you what she has done for thousands who have gone before. Keep her confidence and do not betray her by carelessness or misdeed. Do these things and the ship will bring us to the mouth of the Clyde on Tuesday next—so help us God."*

The RMS *Queen Mary* astounded individuals with her unrelenting success as a luxury ocean liner. Without a doubt, she is known as being one of the most magnificent and fastest vessels ever designed for the seas. Cunard board chairman Sir Percy Bates confidently described the *Mary* as a ship of peace. As a converted warship, she continued to prove her success on the oceans.

The sheer size of the *Queen Mary* and *Queen Elizabeth* meant that both liners could transport more than twice as many servicemen as any other British vessel. Furthermore, this permitted other ships to carry equipment and supplies so combat units could be dispatched to crucial locations. The *Mary's* ability to carry and run whole divisions on urgent deliveries definitely helped to turn the tides of war. Her efforts helped lessen the duration of the war and undoubtedly saved numerous lives.

The outbreak of World War II on September 3, 1939, placed the *Queen Mary* in a precarious situation. In Europe, war was imminent, and in August, the clouds of war became even darker. The British Admiralty ordered all United Kingdom merchant liners to not partake in any regular trade routes.

Stationed just across from the French liner *Normandie*, the *Mary* patiently awaited her next call of duty. According to Cunard, she was supposed to go back to Southampton. However, the Admiralty cancelled all of her future voyages in order to help protect her from threats and disaster at sea. Many of the liner's present staff was sent home to England on smaller ships.

On September 2, 1939, the *Queen Mary's* master acknowledged a coded Admiralty message ordering him to place her on full war alert. Furthermore, he was mandated to enforce all precautions necessary to ensure the liner's safety. The passengers on the last commercial voyage were filled with anxiety due to the pending Second World War. In the morning hours of September 5, 1939, the stately ship arrived safely into New York. The following day, the Cunard-White Star office, located in New York, cancelled all of the *Mary's* future scheduled commercial voyages.

Thus, the *Queen Mary* started her six-month idle at Pier 90 in New York's harbor. Initially, it was thought that the idled ships were enemy aircraft targets. Cunard's newest 83,000-ton liner, *Queen Elizabeth*, was also thought to be one of those targets. The *Queen Mary* and her sister ship would lie in port adjacent to one another. The ones that were enemy targets feared that if they left port they would become sitting ducks as German submarines patiently awaited off the U.S. Atlantic coast. American intelligence reports indicated that Nazi agents in America planned to blow up the *Mary* or set her on fire.

Thus, the United States wondered what the best course of action would be for the ships waiting out the war in New York. New York Police made sure that the vessels were safely guarded from their enemy. Any ship of a substantial size was a constant target for Nazi Germany.

However, a liner is more vulnerable to disasters such as fire and deterioration as it lies quietly in a port.

As the months carried on, the idea to requisition the sister ships for wartime roles gained valuable attention. The sheer magnitude of the big ships would prove beneficial when the constant need to transport several troops would arise. The big liners were also better equipped to travel a variety of routes that some of the smaller liners could not manage. Most advantageous was the fact that the two giant Cunarders were extremely fast on the seas and would prove to be more economically savvy than the smaller liners. After much debate, it was decided to utilize the giant Cunard ships as troop transport vehicles. The *Queen Mary* was the first ship chosen to partake in the new wartime task.

President Roosevelt made a suggestion to Secretary of State Sumner Welles that the United States take ownership over the *Normandie* and the *Queen Mary* to be applied to war debts. Roosevelt's plan stipulated that the liners would fly the neutral American flag and be utilized mainly for transportation of marooned American soldiers from European war zones. France and England rejected Roosevelt's plan.

The British Ministry of War Transport requisitioned the *Mary* for wartime duties. Approximately 500 crew members from the *Antonia* were sent to help prepare the big liner for her war career. One of the biggest tasks for preparing the *Mary* for war was the generous slap of grey paint that she received in order to help camouflage her from the enemy. Her lettering was entirely removed and her lights extinguished. The *Mary's* luxury appointments had to be removed and stored for the entire war period. The ship's delicate fittings were taken apart and crated. Her chairs, sofas, pylon lamps, and glass bookcases were taken off the ship. Six miles of Wilton carpeting and 220 cases of bone china, crystal glassware, and silver settings were stored in New York. The majority of her lifeboats were tested and examined. Lookouts were positioned on the *Mary* to scrutinize the horizon for any sight of the enemy.

Preparing the Cunard Queens for war was an interwoven tapestry of many intricate responsibilities. While moored in the Boston Navy Yard, U.S. Navy surveyors and architects collaborated with the two liners' construction drawings in order to figure out which locations on the ships would be used for certain tasks, such as berthing, bathing, and eating. All in all, a systematic routine of accessing the vessels' facilities had to be coordinated. The *Queen Mary's* troopship conversion process proved to be more complicated than the work done on her sister ship.

In order to accommodate thousands of military troops, certain alterations and refitting were required prior to war service. Within a period of two weeks, an excess of bunks, toilets, showers, sinks, kitchen facilities, and stores were mainstreamed on the ship. The Observation Lounge was changed into a maze of five-tier bunks. The Midship's Bar on the portside of Promenade Deck contained 250 bunks with a small air space between each bunk. When the *Mary* carried 15,000 troops, there still was not enough space to accommodate each sleeping soldier.

The *Queen Mary* acquired many types of armament on her Sun Deck. Thirty-three guns, twelve rocket launchers, a range finder, and a central gun control house could be visible on her top decks. A four-inch gun was mounted on the liner's fantail. She also received a de-gaussing girdle and a band of wire strapped to the hull and charged by an electric current to prevent the hull from receiving damage from magnetic mines. Vintage Lewis and Vickers machine guns used in World War I were installed to detect any reconnaissance plane that the liner could possibly come into contact with. In order to further help evade submarine attacks, the *Mary* received the installation of an ASDIC underwater sound detection system.

The Cunard Queens' primary mode of defense was evident in their intense sailing speed. Thus, proper care and upkeep of the boilers and engines took precedence. The Engineering Department took control of this task, where the Staff Chief Engineer and Chief Engineer oversaw the duties of the stokers, boiler tenders, and machinists. It must be said that neither sister ship endured any engine or boiler failure during World War II.

Other departments were crucial to the overall function of the sister ships during World War II. The Purser's Department continuously worked all hours each day leading up to the vessels' departure from New

York City and until she arrived in Gourock. The Deck Department consisted of the seamen and officers who were outfitted with the task of sailing the ships and delivering the thousands of servicemen to their destinations. The Military Police took care of any critical discipline issues. The Military Staff held the responsibility for delivering the captain's daily orders as well as implemented the standing orders. This staff also made sure that the troops abided the security, blackout regulations, and followed through with their assigned work task.

On March 19, 1940, the master of the *Mary* was called upon to the line's offices in Manhattan. It was instructed that the *Mary* had to be ready to set sail within two days. Cunard officials maintained their usual silence as to the vessel's travel destinations. However, on land, it was still reported that the *Mary* was bound for Australia and New Zealand to carry Anzacs to European concentration points. The liner would travel to Sydney via Trinidad, Cape Town, and the western port of Freemantle.

Engineers had been slowly warming the liner's engines for a day, which was a standard procedure prior to any voyage. The three international code signals were flown from her gaff, which identified her to American ships in domestic waters. The *Mary* flew the American flag from her foremast. Lifeboats were tested and heavy security patrolled Pier 90. British Intelligence retained hope that Axis agents would take notice of the *Mauretania's* departure from New York on March 20, 1940, while missing the *Mary's* departure the following day.

March 21, 1940, marked the day that the *Queen Mary* slowly approached the sea as a wartime vessel. She was en route to Australia to undergo troopship out-fittings. It was just after 8:00 a.m. on a New York City morning when the tugs directed the *Mary* downstream. Many commuters clapped their hands and waved at the massive liner as she traveled toward the ocean. Once she reached her open element, she sailed in a southward direction toward Cape Town, South Africa. She made a brief stop in Cape Town to refuel prior to sailing to Sydney, Australia.

"It was a strange voyage," recalled Harold Blakely, a crewman who managed the refrigeration aboard the ship. Blakely continued to state that, "none of us really knew where the ship was bound, but in the warm waters of the South Atlantic we were sure we were not headed for England. When we reached the Cape we knew we were headed for the Pacific."

Once in Sydney, the *Queen Mary* was still being requisitioned for war. While docked at the Cockatoo Docks and Engineering Company, the majority of the shops in the vessel's main hall were changed into military offices. Welded hinged steel blast shutters were placed over the bridge windows. The *Mary* had to make room for a massive amount of living cargo: 5,500 Australian soldiers and their military gear as well as 900 crew members. At last, the stately liner was converted into one of the biggest and quickest troop ships in the entire world.

On March 23, 1940, a local New York newspaper relayed that the *Mary* and *Mauretania* would become important links in Canada's huge air training program, transporting troops and supplies between Australia and the western portion of Canada. There was a very small chance that either liner would be attacked on their sailings between Sydney and Vancouver.

On May 4, 1940, the *Mary* left Sydney with other converted Cunard ships. The *Mary* sailed from Australia on a zig-zag course for the Clyde River. This was the beginning of the *Mary's* trooping duties. A few days later, the liner started her 6,000-mile sailing across the Indian Ocean to Cape Town.

Situations in Europe continued to go downhill in 1940. On June 10, 1940, Italy declared war on Britain and France. On June 16, 1940, the *Mary* arrived at the Clyde, delivering Australian men. The Australian soldiers were stationed at Salisbury Plain and then off to Colchester for the imminent German invasion of Britain.

The *Queen Elizabeth* and the *Queen Mary* continued to sail in and out of Sydney, Australia, taking troops to European fronts until the Japanese bombed the Pacific in 1941. Thus, the Australian port was considered dangerous for the two giant liners. When the United States entered the war, New York became the functioning port for the two liners.

During the war, the *Mary* traveled 600,000 miles and delivered 500,000 American soldiers to their destinations. The *Mary* also carried 100,000 British troops as well. Her massive armament of guns was never fired at the enemy. It must be mentioned that the RMS *Queen Mary* never encountered a torpedo as she sailed the seven seas during her wartime duties. Not once under the command of Commodore Sir James G. P. Bisset did the *Mary* see a torpedo or a submarine. The grand liner traveled through war zones time after time and she was never attacked by air. However, imminent danger to the *Queen Mary* occurred in 1943 when she was located just off the British Isles. Apparently, a mysterious explosion sent up geysers of water but did not leave a detrimental mark on the liner.

"Perhaps it was a spent torpedo at the end of the run," declared Commodore Bisset.

Despite never being touched by a torpedo, the *Mary* did encounter near-disaster incidents that could have surely lead to her demise. Adolf Hitler offered a prize of $250,000 to the submarine crew that could sink the massive *Queen Mary*. Additionally, Hitler would reward the submarine's skipper the Iron Cross award with Oak Leaves. No enemy submarine ever came close to sinking the *Mary*.

Furthermore, it was feared that Nazi personnel were in Scotland trying desperately to disclose the *Mary's* voyage schedule and location. British Intelligence was meticulously working day and night to protect the grand liner. Penalties were given to those who divulged any secret information as to the *Mary's* location. Any person that disclosed any secret information regarding the liner was transferred to onshore assignments. The rigorous precautions taken were absolutely necessary for the nation's security as well as the security of the *Mary* herself.

When the *Mary* was scheduled to sail into Trinidad to refuel and stock her provisions, a German submarine quietly came into the Trinidad harbor and sank two ships in idle. On her way over to Trinidad, the *Mary* received an urgent coded message mandating that she change her route. Thus, she changed her course and sailed into port in Brazil.

In June of 1940, the ship's captain had sealed orders, which he opened only after the *Mary* had cleared the Firth of Clyde after her temporary stay to refuel and re-stock. On the ship's next sailing orders, she was ordered to take forces to Singapore. Thus, the *Mary* traveled to Singapore by way of Cape Town and Trincomalee, Ceylon.

The *Mary* was a part of an enormous fleet, which consisted of Canadian Pacific's *Empress of Britain* and *Empress of Canada*; Cunard's *Franconia*; Royal Mail's *Andes*; P & O's *Stratheden* and *Strathaird*; Orient Line's *Orion, Otranto,* and *Ormonde*; Furness-Bermuda's *Monarch of Bermuda*; Gdynia-America's *Batory*; British India's *Aska*; and the Australian armed merchant cruiser *Kanimbla.*

For 41 days, the *Mary* received some upgrades. The barnacles on her hull were taken off. The ship's steering gear and engine room machinery were overhauled. Another generous coat of grey paint was applied to the massive vessel. After her sojourn in Singapore, the ship would travel back to Sydney to resume her trooping schedule.

The *Mary* encountered another crisis at the onset of the Italian invasion of Egypt in September of 1940. The need strictly rose for allied reinforcements due to the possibility that the Suez Canal would be taken. Thus, the *Mary* was refitted and destined for Sydney. The *Mary* arrived back in Sydney on September 25, 1940, and engaged in three more weeks of work specifically done to increase the capacity of troops.

She then started her Indian Ocean shuttle runs. The vessel started transporting troops to the Middle East in October of 1940. For the next several months, the *Queen Mary* crossed the Indian Ocean in a zig-zag motion. To help keep her safe from the enemy, the *Mary* was allowed to travel no further than Bombay or Trincomalee where her fighting men were changed over to smaller liners for the expedition to Suez.

The *Queen Elizabeth* was getting pumped for the start of her trooping duties after the Italian invasion of Egypt took place. The *Mary's* sister was ready to start her war service by the latter part of March in 1941. The two stately sisters traveled together for the very first time in April of 1941, when they transported 10,000 Australian and New Zealand troops to Suez. The remaining months in 1941 were quite busy for the two Queens as they resumed their ferry service between Australia and the Middle East.

The year of 1941 marked a very difficult time for both the *Queen Elizabeth* and her sister ship. Both vessels were designed to specifically travel in the cold North Atlantic region. Thus, air-conditioning was never warranted. As both ships traveled through the Indian Ocean and Red Sea in blistering heat conditions without air-conditioning, the lower decks would reach triple digits.

Thus, many men, including POWs, suffered physical ailments due to the oppressive heat and for being packed in like sardines. Many injured POWs passed away due to the rising temperatures aboard the ship. The extreme heat caused anger and irritability among the troops. Arguments and gang warfare were prevalent. In order to halt fights among the troops, ship security took measures to hose the men down with water. Salt-water showers were also implemented to prevent any insubordination. Tragically, in the summer of 1941, several men succumbed due to heat exhaustion and others came close to death.

By the end of November in 1941, both sister ships had held 80,000 Anzacs that they delivered to Europe and the Middle East. They also held many Italian prisoners of war as they were transferred to internment in Australia. The Cunard Queens' efforts enabled Britain to alleviate her defense lines in North Africa and thus slow the Axis advance toward Egypt.

However, the Japanese attack on Pearl Harbor in December of 1941 destroyed any hopes for the conclusion of World War II. Great Britain's strategic planning was severely changed due to the annihilation of the American Pacific Fleet and Tokyo's pronouncement of war against the allies. Japan's entry into World War II on the Axis side endangered Britain's critical supply line to Australia and the Indian sub-continent. Furthermore, the safety of the ports in Singapore and Hong Kong were compromised and could turn into a death warrant at any moment should Japanese troops travel southward. Additionally, the danger of a Japanese attack on Australia posed a hazardous threat.

The attack on Pearl Harbor had some positive features as well. First off, it caused the United States to enter the war on the allied side. Furthermore, Great Britain and the United States were allied together against Fascism, so the *Queen Mary* would play a vital role in the joint venture. President Franklin D. Roosevelt quietly collaborated with British Prime Minister Winston Churchill on a combined tactic for the defeat of both Nazi tyranny and Japanese development.

The *Queen Mary* was ordered to her familiar port in New York harbor because it was unsafe for the liner to drop anchor in Singapore or any port in the South Pacific or Indian Ocean. Therefore, the *Mary* arrived in the Hudson River on January 12, 1942, after sailing via Cape Town and Trinidad. The grand liner received a gracious welcome as it had been a long time since the *Mary* had been in New York waters.

Tragically, the *Normandie* would never again sail out of New York harbor. The United States government seized her in December of 1941, just after the Japanese attacked Pearl Harbor, and she was transformed into a troopship while she was stationed at her pier. On February 9, 1942, a fire erupted on the *Normandie*. Sadly, firefighting efforts proved unsuccessful at saving the liner and she heeled over and sunk. After World War II ended, the damaged ship was sold to local scrappers at Port Newark, New Jersey. Her remains were destroyed.

As the *Mary* idled in New York's harbor, plans were underway for her future responsibilities in the allied war effort. Winston Churchill and his elder staff members visited the United States at the end of 1941 and both Cunard liners were a hot discussion topic with American war planners.

After much conversation, it was agreed that the United States would supply the majority of soldiers and necessary material needed to stop Japanese advance. Britain would supply the big vessels mandatory for the carrying of troops and their equipment to Australia. Obviously, both the Cunard Queens were chosen to take on this arduous task.

The *Queen Mary* traveled to the Boston Navy Yard on January 27, 1942, and underwent modifications that enabled her troop capacity to be increased from 5,000 to 8,500 individuals. The ship's former first- and third-class pool, Promenade Deck, and ladies' Drawing Room all received sleeping bunks. More toilet facilities were installed and room was spared for housing the extra tons of food and water needed for the additional troop load. The *Mary's* armament was also improved while she was in Boston. She received ten 40-mm cannons in five double mounts positioned fore and aft, twenty-four single-barrel 20-mm cannons, six 3-inch high flow angle guns, and four old-fashioned 2-inch anti-aircraft rocket launchers.

The *Queen Mary's* very first voyage carrying American troops began on the night of February 17, 1942. An excess of 8,000 soldiers touched upon the vessel's decks. It wasn't until the following day that the *Mary* sailed out of Boston and changed her course due south. The route would take the vessel to Sydney by way of Trinidad, Rio de Janeiro, Cape Town, and Freemantle. However, U-boat activity close to Trinidad coerced the *Mary* to redirect to Key West.

It was at Key West that Captain James Bisset took command of the stately Queen. He was taken to the Queen via a Navy tug boat. He was a veteran Cunarder, having devoted 35 years to the company. Two 6,000-ton supply tankers were positioned at each side of the liner. Two U.S. Navy destroyer ships surrounded the *Mary* constantly, helping to ensure her protection against the enemy.

As Bisset approached the massive ship via his tug boat, he echoed the following words:

> *"(She had) the appearance of a great rock set in the middle of the sea. Two U.S. destroyers were patrolling around the anchored ship, on the lookout for U-boats. As we drew nearer, I could make out two tankers, one moored on each side of her, amidships, feeding her with oil-fuel and fresh water. Though the tankers were vessels of 6,000 tons, they were dwarfed by her tremendous bulk. Gazing up at her I felt overawed at the responsibility soon to be mine."*

Once the ship sailed through the Anegada Passage into the Atlantic, two German submarines were also traveling along the same route. Perhaps the distress call from a torpedoed allied oil tanker helped to save the *Queen Mary* from disaster. This close call was one of many the ship would face on her way to Australia.

Later on, a group of German and Italian spies made the discovery of the *Mary's* traveling schedule and transmitted this crucial piece of information to U-boats located off the Brazilian coast. Luckily, allied intelligence intercepted their message. Thus, Bisset was ordered to remove the *Queen Mary* from the Rio many hours ahead of her schedule, which most likely saved her.

It must be mentioned that an oil tanker was sunk as it left the port at the exact same time the *Mary* was originally scheduled to be in port. The Germans thought for sure that they had the *Queen Mary* in the bag and even broadcasted the news of her demise. Avoiding disaster yet another time, the majestic liner started on her 3,300-mile venture to Cape Town on March 8, 1942.

As the *Mary* raced toward South Africa, a life-threatening fire erupted on B Deck right below the ship's bridge section. The liner was 1,500 miles off shore and unescorted. Apparently, the two-hour fire was ignited by a fault in the electrical insulation. Smoke and flames rose to the bridge level. Keeping panic and fear at bay, everyone performed their duties perfectly. Crew members were trained to abandon ship, but fortunately, that never happened.

Later on in her war career, while sailing near Bermuda, the *Queen Mary* spotted seven lifeboats filled with exhausted men. The big liners that served in the war were given orders by the Navy to never stop to help survivors of ship wrecks. Thus, the *Mary* sailed right by the seven lifeboats. However, the ship sent a blinker signal to the lifeboats, which relayed to the stranded men that word of their condition would be sent out via wireless. The following day, the men were effectively picked up by American naval units.

European forces desperately needed reinforcement by the spring of 1942. Thus, both of the Cunard sisters were employed to transport the gigantic transfer of soldiers. The *Mary* began the GI Shuttle program as she brought over the first wave of Americans to Britain. She sailed to Gourock in five days and three hours, carrying 10,755 souls.

In 1942, the Queen liner continued her trooping duties across the oceans. On May 7th, the vessel commenced a 78-day journey of 35,000 miles. It must be mentioned that the *Queen Mary* sailed some of the toughest water terrain on the planet and outsmarted the joint military and naval forces of Germany, Italy, and Japan.

When Winston Churchill was in Washington D.C. in the latter end of 1941, General George C. Marshall asked him if the Cunard Queens could be revised to allot for 15,000 men. Marshall's concern with this plan was that both ships only had enough lifeboats for about half of that capacity. Marshall himself visited the *Mary* to visually examine the situation.

Furthermore, the *Mary's* Staff Captain, Harry Grattidge, was worried that the weight of 15,000 bodies might cause the liner to list and potentially scrape the tip of the Hudson Tunnel as the liner headed toward the sea. Grattidge came up with a solution to the problem. The troops were instructed to be absolutely still until the ship passed through the Hudson Tunnel. Indeed, every single soldier obeyed the task in order to prevent the ship from listing.

The *Queen Mary's* physical modification necessary to account for 15,000 soldiers was delayed due to the developments that took place in the Middle East. The German-Italian assault of Libya at the end of March of 1942 coerced London to transfer large amounts of troops from Great Britain to the Mediterranean. The *Queen Mary* was desperately needed to carry the soldiers from New York to Scotland. British soldiers would come aboard the ship in Scotland and were transferred to Suez. Then the liner was modified to hold an entire division of 15,000 men.

May 11, 1942, marked the first date in history that any large vessel carried 10,000 people. It was also the very first time that the *Mary* held American soldiers who were bound for the British Isles. The *Queen Mary* eventually reached New York on July 21, 1942, to undergo modifications to increase troop capacity.

On August 1, 1942, the *Queen Mary* was ready to start her journey carrying 15,000 troops per voyage. The whole First U.S. Armored Division, totaling 15,125 men, came aboard the ship during the night. General Marshall's plan proved to be effective as each soldier remained completely still as the *Mary* cleared the Hudson Tunnel. For the next few months, the vessel made voyages between New York and Gourock.

THE HMS CURACOA DISASTER

The *Queen Mary* was delivering 10,000 American troops of the 29th Infantry Division and a total crew of 850-plus when she traveled off the coast of Donegal, Ireland, prior to sailing to the Clyde estuary to pick up the HMS *Curacoa*, a British Navy escort ship. This smaller vessel was a "C" class light cruiser, having served during World War I. In July of 1916, she was built in the Pembroke Dockyard located in Wales. Her launch took place on May 5, 1917. Upon commissioning, she united with the Harwich Force, an assortment of cruisers and destroyers tasked with the role of protecting the English Channel's shipping. Captain John Wilfred Boutwood commanded the cruiser with a total crew of 439 men at the time of its demise.

The time was approximately 10:00 a.m. when the light cruiser sailed about five miles ahead of the *Mary*. Right away, a sequence of events and circumstances began to take shape and form, ultimately leading up to the heartbreaking catastrophe. For one, the smaller vessel could only travel at a maximum speed of 25 knots, 1.5 knots slower than her larger convoy. Their differing sailing rates led up to the *Queen Mary* surpassing her escort ship in just four short hours, leaving Captain Boutwood without the critical information necessary to locate the larger liner at the exact time the two ships collided. As the ships eventually inched closer to one another, Boutwood did not know when the *Mary* was due to alter course.

Known as one of the most tragic maritime misfortunes, the *Curacoa* disaster should have never occurred. The visibility was impeccable and both ship captains seemed to be visually aware of each other's whereabouts, speed, movement, and course. However, both Captain Boutwood and the *Mary's* Commodore Giles Gordon Illingworth should have better communicated with each other. If they had, chances are that the looming tragedy would have been surely averted.

Furthermore, just one hour prior to the fatal incident, the *Mary's* compass was off by a miniscule 2 degrees, a tiny number that led up to a sizeable disaster. The course should have been 108 degrees; however, it was only 106. At this point, the two ships kept separating from each other, causing Boutwood to alter the *Curacoa's* course, initially to 105 degrees, then to 100 degrees, initiating the inevitable collision.

The only damage that the *Queen Mary* sustained in World War II occurred on that fateful day of October 2, 1942, at 2:14 p.m. in the afternoon. The *Mary* was on a Number 8 zig-zag course to avoid the enemy when she hit and sliced in half the HMS *Curacoa,* who was escorting her along with six other destroyers. The grand liner struck the 4,000-plus-ton light cruiser and the sheer weight of the *Mary* was enough to turn the smaller vessel around, enabling the *Mary's* bow to slice right through the cruiser.

A bump was felt as the massive ship ran into the light cruiser.

"Was that a bomb?" Illingworth pondered.

"No sir," said the helmsman. "We hit the cruiser."

Terrified and screaming sailors leaped from the sinking vessel into the sea. The American troops aboard the *Mary* threw lifebelts and anything else that would float. The radio officer was instructed to contact the other escorting destroyers to help in picking up surviving men. The *Mary's* commodore had an urgent and difficult decision to make at the time of the incident. He pondered whether to ignore Navy orders and stop to rescue the surviving men of the cruiser and subsequently risk the thousands of living individuals aboard the massive ship or sail right beyond the survivors. Illingworth chose the latter.

The light cruiser met the bottom of the ocean within five minutes of being struck by the *Mary*. Only a fraction of the cruiser's men survived, including Captain Boutwood. As any dutiful captain, Boutwood remained with his sinking vessel up until the end. When the bridge went under, he floated away. Many of the surviving sailors watched in horror as their friends and shipmates went down with their cruiser. Tragically, over 300 British sailors aboard the smaller vessel died as a result of the accident. 101 survived. The recovered remains of the *Curacoa's* sailors are buried in one of two cemeteries on the Isle of Skye and Arisaig's mainland.

Commodore Illingworth was responsible for the difficult task of notifying the Admiralty of the tragedy. At 2:20, he delivered a signal to White-hall: "HMS CURACOA RAMMED AND SUNK BY QUEEN MARY IN POSITION 55.50N 08.38W. QUEEN MARY DAMAGED FORWARD. SPEED TEN KNOTS." A short while later, Illingworth sent to HMS *Bulldog*: "IT WOULD APPEAR THAT CURACOA ATTEMPTED TO CROSS MY BOWS WHEN COLLISION OCCURRED. AM REDUCING SPEED TO ASCERTAIN EXTENT OF DAMAGE AND HAVE CEASED ZIGZAG. WILL KEEP INFORMED." The *Queen Mary's* bulkhead only sustained minor damage and as a result, her speed shortly after increased to thirteen knots.

Phillip Levin of New York was a sergeant major aboard the liner at the time of the collision. In his own words, this is what he had to say:

> *"I can recount that at 2:07 p.m. in the afternoon the lookout raised alarm that a suspected U-boat was spotted on the port bow ahead. In response, the* Mary *wheeled to starboard. In the meanwhile, the* Curacoa *was answering the submarine alert and sped to port. She cut in front of the liner's bow and because of a split-second miscalculation, was trampled by the larger vessel. I was in the office on the Main Deck at the time and felt only the slightest rattle and vibration. Actually, it seemed quite normal. But word of the collision spread quickly, like wildfire throughout the ship. I raced to the open, upper decks and looked aft to see the two halves of the* Curacoa *drifting in our wake and then rather quickly sinking. I could see the drowning sailors and also the pick-up of the survivors by the other escort ships, which had raced to the scene."*

Commodore Illingworth watched horrifyingly as the cruiser quickly slumped below the waters. He mandated that Staff Captain Harry Grattidge go below deck to scrutinize the damage to the *Queen Mary*. He then ordered that the liner's speed be reduced to ten knots. Grattidge relayed:

> *"The speed was still on the ship when I reached the forepeak. By the light of a torch I could see the water racing in and out of the forepeak, a great column of it forming a kind of cushion from the collision bulkhead, the watertight reinforced steel wall that rises from the very bottom of the ship to the main deck. If that bulkhead were weakened I did not like to think of the* Mary's *chances. I sweated through my silent inspection. But finally, not a crack. Not a break. I turned to the Bosun and the carpenter: 'Get every length of wood you can find, Bosun. Get it down here and strengthen that collision bulkhead as much as you possibly can. I'll report to the Captain.' ... I was sick at what we had done, yet I marveled, too, at the strange and terrible impregnability of the* Queen Mary. *It came home to me that she had no equal anywhere in the Atlantic, perhaps not anywhere in the world."*

It must be mentioned that the *Queen Mary* came close to seeing her end when she hit the HMS *Curacoa.* You see, the light cruiser carried depth charges that were situated on her aft end. If the *Queen Mary* had hit the cruiser at that exact angle, the explosions could have damaged her bow, which would most likely have caused her demise.

Was this sheer luck? Or was she being guided to safety?

The hole in the *Mary's* bow was filled with cement and she traveled to Boston to receive a new bow plate. The reason she was not repaired in dry-dock was due to the continuous threat of enemy air invasions of the ships in the Clyde shipyards.

In 1947, the British Court found that the HMS *Curacoa* was at fault for the accident that occurred back in October of 1942. Thus, the *Mary* was not to blame for the cause of the tragic accident. The Admiralty Division of the High Court agreed that the tragic catastrophe was caused by the negligence of the cruiser. Presiding Judge Sir Gonne St. Clair Pilcher released the action that was brought by the Admiralty against the Cunard-White Star Line. He also instructed the Admiralty to take care of the trial costs. Later, it was found that the cruiser was responsible for two-thirds and the *Queen Mary* one-third. Regardless of the outcome, both Captain Boutwood and Commodore Illingworth endured heartache and sorrow for the remainder of their years.

The Queen Mary Marches Forward

Perhaps the most anxiety-provoking incident during the *Queen Mary's* sailing career had nothing to do with war at all but rather the angst of the sea. During a December 1943 voyage while carrying an excess of 10,000 GIs, the ship was hit by a monstrous rogue wave that ultimately could have caused her to capsize. The sheer weight of the wave made her heel over to starboard, initialing the roll that many thought would end her days. She continued to roll, stopping at about 52 degrees until she miraculously recovered to a normal keel. Even though injuries prevailed, not one person succumbed as a result of the event. It was later discovered that if the ship had rolled just three more degrees, both she and her living souls would have surrendered to the sea. As I mentioned earlier in this book, I do believe that the *Queen Mary* has been protected since the dawning of her days.

The war situation on all fronts had radically changed its course by the spring of 1943. The United States Navy emotionally recovered from the attack on Pearl Harbor and was ready to strike back at the opposing Japanese. American, Anzac, and British troops thwarted the Japanese, and it was clear that the enemy was no longer in charge of the Pacific war. Operation Overlord was the allies' ultimate plan in defeating Nazi Germany. This operation was the code name for the Battle of Normandy, the allied effort which initiated the successful invasion of Western European territory. It commenced on a day that we all remember: June 6, 1944. It lasted through August 25th of that same year. In order to carry out the arduous plan, the necessary soldiers and equipment were needed to be delivered to the battle areas.

Operation Bolero was a plan developed to makeshift the British Isles into a gigantic supply repository and staging arena where the cross-Channel attack would be set forth. The Cunard Queens retained their New York to Britain shuttle services where they could both transfer a full division of soldiers to Britain in just a few days. The sister Queens were accompanied by the other vessels *Aquitania*, *Mauretania*, *Nieuw Amsterdam,* and *Ile de France* and would allocate themselves entirely to Operation Bolero. All six of these ships were together known as "The Monsters" due to their size.

The year of 1943 marked the busiest time period for the *Queen Mary* during the war. As we know, she started her adventures as a GI shuttle on June 1, 1943. For the next 23 months, she resumed a travel schedule that mirrored that of her luxury days. In fact, from June 1943 to April 1945, the *Mary* continued on with her GI shuttle services. In that time, she managed to travel an excess of 180,000 miles and carried approximately 340,000 American and Canadian men to the United Kingdom.

Since the *Queen Elizabeth* and *Queen Mary* were prioritized ships during the war, the arduous task of supplying them was an ongoing event in and of itself. Even though the war caused shortages, both Queens were loaded with all necessary supplies. Every day, the *Mary* burned a total number of 1,000 tons of fuel. She typically traveled with 8,000 tons of fuel. Furthermore, she also contained 6,500 tons of fresh water, which was gobbled up at a rate of 700-800 tons per day. Drinking water was filtered and chlorinated and seawater was utilized in the bathrooms. Additionally, during the GI shuttle program, both sister ships had an operations crew that consisted of 800-900 individuals. The shopping list for an excess of 15,000 people was quite enormous, to say the least.

It must be mentioned that the *Normandie's* demise proved to be a serious disadvantage to the allied cause. The French liner would have been utilized for the Atlantic GI troop shuttle along with the sister Queens. There was a consistent undercurrent of anxiety among the troops for fear of a German attack. During this time, the *Mary's* armament was increased as well. Fifty American sailors were placed on the ship as gunners. The only time that they fired a weapon was during practice routines or for confidence and entertainment purposes.

Entertainment was provided for the troops to ease some of the anxiety and fear. Typically, two films were shown a day as well as occasional concerts. Troops also had the option of listening to portable record players and engrossing themselves in games, puzzles, and playing cards.

Troop training for embarkation purposes was provided at Camp Kilmer. The camp was located in New Jersey. Huge wooden simulation ships were developed, so troops could acquire adequate training. Interestingly, each practice event was filmed and shown in the camp's theatre. This way, the servicemen could review any mistakes that were made. The majority of embarkations were successfully finished within 12 hours.

The night prior to any sailing, the fighting men arrived by ferry or bus. Once the troops came aboard the vessel, each man was given an assignment card or a coded metal disc, which would display his area for eating, sleeping, and location for drills. The following dawn, troops were called upon for instructional drills designed to highlight the liner's routine, emergency drills, black-outs, air-raid procedures, and abandon-ship methods. It was mandated that all men have possession of their lifejacket at all times. Men who failed to do so would receive consequences. Disembarking the vessels was just as intricate as the boarding process. Once the servicemen left the ship, a substantial cleaning and maintenance unit came aboard to commence the process of preparing for the next batch of troops.

The ship usually departed in the early morning hours or late evening hours to take advantage of the maximum high tide and to conceal the liner's movements from view. The *Mary* never sailed on the same route twice. In fact, her master did not even know the precise course she would take until the ship left port. After traveling past the Ambrose Lightship, the master was permitted to open the sealed orders that were handed to him at the Shipping Office. He then gave the orders to the navigator and instructed the helmsman to start the zig-zag sailing motions.

Furthermore, the *Mary's* captain, on each sailing, addressed the men on the imperative subject of the liner's survival.

There were certain procedures, routines, and orders put in place for the *Mary* during her GI shuttle days. A standard voyage began with a senior officer's meeting at the Allied Combined Shipping Operations Office located in New York City. During this meeting, the *Mary's* master consulted with representatives from the British Ministry of War Transport, the U.S. Army Transportation Corps., the New York Port of Embarkation, and the United States, Canadian, and Royal Navies. The meeting ironed out information on routing, troop loading, and escort measures. Intelligence officers relayed the current whereabouts of enemy liners.

Prior to any shuttle departure, the fuel and water lines were connected to the ship's tanks. Supplies and ammunition were brought on the liner. This signaled the beginning of another GI shuttle cycle. On a standard GI shuttle journey, the *Mary's* company contained 850-910 men, separated into four working sections. The sections consisted of the Deck Department, the Engine Department, the Catering and Purser's Department, and the Permanent Military Staff. Each section had to fulfill a pre-departure and "at sea" function.

The *Queen Mary* was one of the largest liners ever used as a troop transport ship. She was more than 30,000 tons heavier than the *Leviathan*, which carried 100,000 men during World War I. In total, the *Queen Mary* and her sister ship, the *Queen Elizabeth*, carried 1,243,538 soldiers.

The *Mary's* standing orders and the daily orders controlled life aboard the vessel. Some of the vital aspects discussed consisted of emergency procedures regarding air attacks, the occurrence of fire on the ship, and security and safety measures. In order to avoid human traffic problems on the ship, the joint British-American military command that traveled on the ship developed a section system in order to avoid problems caused by the traffic of 15,000 soldiers.

The section system gave each man a ticket that told him where to sleep and where to eat. Each of the three sections was designated as blue, red, or white. Each outlined area also had a canteen, which also helped to remedy the hunger situation that came in between meal times. It was vitally important that each soldier stay in his appointed section. For example, a man with a blue button was not allowed to enter a red-button area.

The red area went from the ship's bow to the number three stairway, not including the Sun Deck. The white section area included all space between the number three and four stairways, including the whole Sun Deck. The blue section area had the rest of the ship from the number four stairway to the liner's stern.

There were only two main meals a day. Breakfast was from 6:00 a.m. to 10:00 a.m. Dinner was from 3:00 p.m. to 7:00 p.m. It was mandatory that each man eat at his appointed time. Meals were served in six sessions lasting 45 minutes in length.

Out of 15,000 individuals, 7,400 were known as double bunkers, which meant that 3,700 men slept in a bunk one night and on the deck the following night. This sleeping arrangement was accomplished via lottery. The sleeping bunks rose in tiers through every available spot on the liner. Additionally, every other public facility not utilized was equipped with rows upon rows of bunks. Many men felt that the ones who slept topside had a more feasible chance to escape during a torpedo attack. Thus, sleeping below deck caused a lot of warranted anxiety for the courageous men.

In order to ensure maximum comfort for each man on board, it was absolutely necessary that each man comply with the regulations. Each soldier was mandated to attend a boat drill at least once a day. Every soldier who was not on current inspection duty was required to gather on the decks where they were individually inspected. Other squads also inspected the bunk rooms and cabins.

The GIs boarded the liner the day prior to sailing, bringing with them their duffel bag, rifle, helmet, canteen, cartridge belt, field pack, and any chosen personal items. When the men were aboard, they were educated on the magnitude of the artwork, murals, and fine woods on the *Mary*. The troops were instructed to not damage or vandalize the artwork onboard the ship. A 750-foot long teakwood railing on the

Promenade Deck was supplied so the troops could carve their initials or names. Suffice to say, the one-third mile railing received names, nicknames, initials, girlfriend's names, dates, and the carving of a naked female.

It must be said that no artwork or mural was spoiled by the troops, which was a testament to the obedience of the servicemen. When the *Queen Mary* returned to her luxury travel years, a six-foot portion of the railing was detached and delivered to the U.S. Army Archives for the purpose of a wartime memento. "They are a grand bunch of boys. It makes you proud you are an American," recalled Col. Dallas D. Dennis of San Francisco, an American infantry colonel who was a transport commander of the *Mary*.

A traffic control system had to be implemented aboard the ship to ensure both functionality and individual safety while aboard the vessel. For example, the starboard thoroughfares were utilized for forward movement, whereas the portside thoroughfares were restricted to traveling aft. Once the servicemen were safely aboard, they were handed a copy of the standing orders and daily orders. The latter was positioned in various areas of the ship and revealed the work assignments, religious services, provided leisure activities, and news concerning war developments. The daily orders' last words were always "ignorance of these regulations will not be accepted as an excuse in any case of breach of discipline."

There was one regulation in particular that the troops consistently disregarded: the ban on gambling aboard the ship. To pass the time away, soldiers saturated themselves in games of poker, blackjack, and craps. Many of the men literally spent most of their time at gaming tables. Movies were shown throughout the day and lectures about various topics were given.

The amount of food that was carried on the *Mary* during each voyage was extensive to say the least. Per each voyage, the liner had a food shopping list of 550,000 pounds. For each wartime voyage, the *Mary* carried 76,400 pounds of flour and cereal, 21,400 pounds of bacon and ham, 155,000 pounds of meat and poultry, 124,300 pounds of potatoes, and thousands of pounds of cheese, jams, fruits, tea, coffee, sugar, butter, and eggs. The troop dining facilities were jam-packed with noisy men who rushed to eat their meal in the allotted 45-minute time frame for each meal sitting. Sadly, many of the consumed meals did not get digested since many of the men suffered from seasickness while aboard. Stabilizers were not yet attached to the *Mary*; thus, she often rolled and pitched with the often-violent weather on the North Atlantic.

During World War II, the *Queen Mary* carried well-known passengers. The liner carried Prime Minister Sir Winston Churchill on three distinct Atlantic crossings. The legions of dogfaces, airmen, corpsmen, and nurses of the American armed forces were among the most important passengers carried on the *Mary* during the war.

During the duration of World War II, the sister Queens carried many celebrities as well. Some of these noted people included Fred Astaire, Bob Hope, Katherine Cornell, Sir Thomas Beecham, Bing Crosby, Douglas Fairbanks, Mickey Rooney, Alexander Korda, and Basil Dean. However, the most famous and influential civilian passenger during those six long arduous years was Winston Churchill. He chose to travel on the *Mary* because of her quick speeds, providing comfort and security. Churchill's first wartime sailing took place in May of 1943.

The Churchill party was inclusive of 150 individuals. In fact, the ship's Main Deck was restricted to Churchill and his entourage. The suite that he occupied was generously refurnished to his liking. Additionally, a dining room, conference room, and a map room were located next to Churchill's suite. A communications room assisted the prime minister with communication with both sides of the Atlantic. Furthermore, the *Mary's* wartime "dry laws" were lifted while the Churchill party was onboard the ship.

SUBSIDIARY WAR ROLES

The *Queen Mary* was an invaluable vehicle for transporting troops throughout the duration of World War II. She also had the auxiliary responsibilities of carrying prisoners-of-war and injured soldiers simply because she could hold more individuals than any other liner. Once the Cunard Queens carried injured or disabled men to the United States and Canada, specific procedures for their care had to be implemented. The spring of 1941 marked the first time the *Mary* carried any prisoner, and for the next 12 months, she consistently held an approximate amount of 2,000 German and Italian soldiers on her return sailings to Australia. She carried prisoners-of-war from the Mediterranean and Middle Eastern war zones to detention facilities.

In fact, everlasting POW camps were instituted in South Africa and Australia. The Cunard Queens carried approximately 170,000 POWs to Canada and the United States by the end of 1943. This statistic continued to rise in 1944 after the wake of the D-Day invasion. As 1943 came to a close, each ship held an average of 5,000 POWs on each westbound voyage.

Certain areas of the *Queen Mary* had to be modified in order to make way for the prisoners. The areas that would retain the prisoners were cleared of any items that could be utilized as a weapon and supplementary locks were placed on certain doors. Barbed-wire encircled their dining facilities and exercise rooms. Other efforts were put in place to prevent the prisoners from escaping or vandalizing parts of the ship.

When the United States stepped into World War II, the *Queen Mary* started transporting prisoners-of-war to New York. During each of her sojourns at New York's port, the liner disembarked thousands of prisoners destined for detention camps in Canada and America.

An excess of 55,000 German troops were sent to America in September of 1944. Many of these prisoners-of-war received their first visual image of the United States via the *Mary's* many portholes. By the end of 1944, the grand liner transported an average of 5,000 prisoners to New York on her return voyages from Scotland.

The *Queen Mary* was also hired to transport injured troops to hospitals in the United Kingdom. After the invasion of France, every single ship that was competent enough to carry passengers was ordered into medical service. In 1944, there was an urgent need for medical evacuation liners. The sister Queens seemed more capable of holding many more injured persons than other liners. Eventually, both Cunard sisters were refitted with better medical amenities.

The modifications on the *Queen Mary* consisted of the installation of 1,084 single- and double-tier hospital beds. A small laboratory was added next to the Promenade Deck. Many large staterooms were converted to exercise and recreation facilities. Main Deck consisted of separate kitchens that were used specifically to make meals for the patients who adhered to a special diet. To put these modifications to work, the British-American medical personnel was increased.

In the spring of 1945, the allies were starting to defeat Germany. The Reich's falling rendered future GI shuttle services unnecessary. Therefore, the *Mary's* next trooping voyage was halted and she became dry-docked in April of 1945. On May 7, 1945, Nazi Germany finally surrendered, ultimately ending World War II. In September, the *Queen Mary* returned to Southampton, which concluded her final voyage as a troopship.

Germany's surrender brought on victory for the allied forces. The weeks after the fall of Nazism, people from Europe to California celebrated and saturated themselves in relief. However, allied forces were still competing in battles in the Pacific and Far East. One of the biggest invasions still yet to come was the last physical attack on Japan. Thus, hundreds of men needed to be re-deployed from Europe to the Pacific. Needless to say, the Cunard Queens and the *Aquitania* were called upon for the plan.

The *Queen Mary* started her repatriation schedule in the morning hours of June 5, 1945, as she traveled from New York to Gourock, the same route she had taken for two years. However, there was one distinct difference this time: for the first time since 1939, the *Mary* was able to voyage in a straight line without the mandatory zig-zag course necessary to avoid enemy forces. The ship's lights shined brightly at night and the black was removed from her many portholes.

On June 10th, the *Mary* successfully arrived in Gourock and was proudly welcomed. Later on in the month of June, the *Mary* returned to New York to receive an over-excited and energetic welcome. As the gigantic liner sailed into view, whistles blew, horns honked, and bells rang. It was the loudest welcome New York gave the *Mary* since her maiden voyage. "We docked at 3:10 p.m., and the troops began disembarking immediately, while bands played and the crowds cheered, laughed, and wept with joy at the heroes' homecoming," relayed the captain.

THE END OF WORLD WAR II

In June of 1945, the *Queen Mary* received a grand and boisterous welcome as she approached the New York harbor with 14,526 troops. Army and Navy individuals crowded the decks of the *Mary* as she steamed into her familiar New York port. Sailing on her decks was the largest single contingent of American fighting men and women to come home from the war. Individuals could be spotted from every porthole on the ship. These American troops consisted of men of the Air Forces, dogfaces of Europe's million foxholes, Wacs, Medical Corps men, naval officers, Army nurses, and seamen. The troops were bound for furloughs at home and the majority of them were destined for further fighting in other areas of the world.

The New York Port of Embarkation, under the provisional command of Col. Hans Ottzenn, sent three of its own vessels down the bay to greet the *Mary* beyond the Narrows. The *Mary* was welcomed by tugs, freighters, Coast Guard and Army vessels, ferryboats, and a water-spouting municipal fireboat. In Brooklyn, rows of cars lined Shore Road. Many people hurried to the Staten Island waterfront to watch and cheer the individuals onboard the *Mary*.

During the war years, the RMS *Queen Mary* retained a world record unparalleled by any other vessel in the world. The stately liner delivered 810,730 troops to Europe and traveled 569,943 nautical miles during the length of battle.

In May of 1946, the *Queen Mary* completed her wartime duties for the United States Army and carried Canadian war brides to Halifax. She operated between the Canadian port and New York solely as a commercial carrier. The U.S. War Department mandated that the *Mary* bring American troops back home on westward voyages. In doing so, she carried the injured, prisoners-of-war, and the many proud troops who were coming back home.

In July of 1946, it was the very last time that the *Mary* would anchor in the Clyde as a war ship. She then joined her sister ship on the very first debut of a two-ship transatlantic schedule. Thus, Cunard profited very well after World War II as thousands of passengers secured a spot on one of the two grand liners.

We must give thanks to the *Mary* and her sister liner for safely transporting the lives of the young, innocent men about to embark on the tedious times of battle. Just as important, we must also commend the servicemen as they were willing to sacrifice themselves for the betterment of the world. We must commemorate these brave and courageous men and women, for their efforts, in tandem with the ships that carried them, gave peace back to the nations. We owe them eternal gratitude.

REPATRIATION AND WAR BRIDE SERVICE

For six months, the sister Queens continued with their repatriation responsibilities. Plans were in effect to transport the wives of American soldiers to the United States at no cost. The first shipment of brides to the United States was expected to sail by the end of January in 1946. The Army relayed that 26,866 wives and children were listed up to December 20th. Eight to ten passenger liners and several United States Army hospital ships were utilized for transporting the wives. As for the *Queen Mary*, she was rumored to have transported 30 percent of all British wives who had married American soldiers. They were allowed to travel to the United States under a particular government plan.

Prior to her conversion back to a luxury liner, the *Mary* was responsible for carrying war brides and their children to designated areas. All British wives and children were sent through the United States Army port at Southampton. Three reception areas were designated and being developed at Tidworth, Perham Down, and in Bournemouth Hotel.

In February of 1946, the *Mary* docked in New York harbor after completing a historic trip across the Atlantic. One-thousand people came to witness the liner sail into her port. The brides hailed from various portions of the British Isles. They were bound for all of the fifty states in America. The average age of the brides was 22 years old. Four-hundred-and-four women on this voyage were pregnant. In total, the *Mary* carried approximately 22,000 war brides and children.

The *Mary* made its way to Pier 90, being escorted by twelve Army tugs. The music of Brahms' "Lullaby" could be heard throughout the area. The 378th Army Service Forces band from Fort Hamilton serenaded the wives and children with "Rock-a-bye-Baby." The wives appreciated their welcome and could be seen applauding and waving their colorful kerchiefs from the ship's portholes and decks. Some wives moved their bodies to the musical rhythms and others clapped the small hands of their babies while the music was playing.

On the way over to New York, the brides traveled on the *Mary* six persons to a stateroom. Any children present would take the place of their mother. The babies traveled in small cribs that were tied to the edge of sleeping bunks. Lifebelts were designed for the children and collapsible chairs were designed to fit inside adult dining chairs. Washing rooms, ironing facilities, and cots were set up on the *Mary*. American Red Cross nurses set up special stations that were located in the lounges.

Many of the women were apprehensive as to the American customs. Thus, on the sailing to New York, they attended daily orientation classes in the liner's elegant lounge. The women intently soaked in as much information as they could pertaining to the various sections of the United States. These orientation classes helped to erase any inaccurate pre-conceived notions or impressions regarding the United States. However, there was still an undercurrent of anxiety that ran through the women.

The one universal concern that the women had was whether the American women would resent them coming over to the United States. Each of the brides was informed that they had a legal right to be in America. They were told that they belonged in America and that it was up to each of them to individually adjust to the new country.

When husbands met their wives, Red Cross personnel held onto the children. Many husbands, upon being interviewed, admitted that they were more nervous to see their wives this time as opposed to their marriage day. While the men waited for the *Mary* to sail to New York, they shared stories from the war and romantic stories as well.

In May of 1946, the *Queen Mary* docked in New York carrying 1,299 American brides and children and 179 civilian passengers, right after dropping off 1,000 Canadian servicemen's families in Halifax. This was the *Mary's* fifth voyage with American families. On September 29, 1946, she arrived at Southampton from Halifax, which signaled the end of her war duties.

Post-War Refitting, Conversion, and Cruise Travel

After World War II, the *Queen Mary* had to undergo extensive refitting and conversions back into a luxury ocean liner. There was some minor damage done to the liner during her days as a warship, which needed to be repaired in order for the ship to be once again a successful vessel. The ship's last war duty was the responsibility of delivering Canadian wives to Halifax. In September of 1946, she was dry-docked for the very first time in an excess of seven years.

At one point, 4,000 individuals repaired machinery and refurbished the many staterooms when the ship was stationed in the Southampton harbor. The entire ship was examined thoroughly to assess the need for repairs or renovations. Furthermore, the carpet and beds had to be changed or renovated. All the decks were placed with new linoleum. Improvements were made to the vessel's air-conditioning and its sprinkler system.

There was much time dedicated to scraping, polishing, and restoring parts of the ship. Ten-thousand pieces of the *Mary's* furniture were stored in New York, Sydney, and England for the duration of World War II and had to be re-installed into the ship. It took nearly a year to convert the liner back into a peacetime vessel. In July of 1947, the *Mary* completed a successful two-day "shake-down" cruise after being refitted as a ocean liner.

"She handles beautifully; she is better than ever; they did a wonderful job," replied Commodore Illingworth on the *Queen Mary's* refitting and conversion process.

In August of 1947, the *Queen Mary* was met with joyful crowds as she sailed into the comfortable waters of New York harbor for the first time since being converted back to a luxury liner. The passengers crowded the Sun Deck to watch as the *Queen Mary* was warmly welcomed. Fireboats saluted the grand liner with streams of spraying water. Other ships saluted the *Mary* by blasting their horns. Planes flying overhead added their own welcomes as well.

The *Queen Mary* and her sister ship, the *Queen Elizabeth*, would see the day when they would pass each other while luxuriously sailing the North Atlantic. At the time, both ships were of equal caliber, which was an unprecedented achievement in the history of Atlantic shipping. Both sister vessels maintained a weekly schedule of traveling from New York to Southampton.

In August of 1966, the RMS *Queen Mary* achieved its fastest post-war trans-Atlantic crossing. She sailed past Bishop Rock, England, at 5:25 p.m. E.D.T., four days and three hours after passing Ambrose Light off New York. She arrived nine hours ahead of schedule, which was faster than her previous post-war record, which was obtained in April of 1962. However, the speed the liner obtained for this record was not nearly as close to the speed it held to achieve the Blue Riband. "I'd go as far as to say the *Mary's* machinery is as good as the day she first left the Clyde and is a credit to the builders," relayed Joseph Parry, the *Mary's* Chief Engineer.

Despite being the most opulent vessel of her time, the RMS *Queen Mary* was not without post-war problems. At one point, the ship went ashore while she was in Cherbourg and had to be patched with many tons of concrete. At sea, passengers were getting quite seasick, so the liner was fitted with stabilizers to prevent her from rolling. She was also held by various strikes. In 1966, the *Mary* was port-bound during the seamen's strike. Later, people complained that the liner vibrated, and it was found that she had 200 tons of barnacles attached to her. Therefore, she was placed into dry-dock to have the barnacles scraped off.

The most devastating problem the *Mary* faced was in 1967. Her captain was told that she and the *Queen Elizabeth* were to be sold, most likely for scrap. All in all, this time period commenced the very beginning of her new era: becoming a popular tourist attraction, museum, and hotel in Long Beach, California.

PHOTO GALLERY

Photo courtesy of Joe Bertoldo

Left: WWII Servicemen on Deck

Right: Anti-Aircraft Gun Emplacements

Queen Mary in Port

Queen Mary During WWII

Queen Mary making her way to Long Beach

Post-war Cunard Queens

Queen Mary Sails On

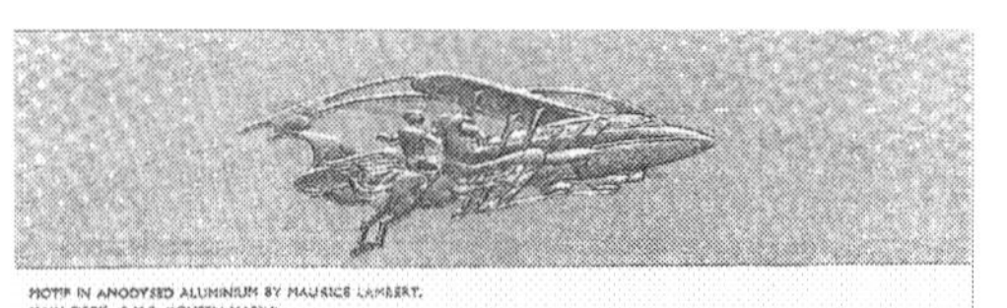

MOTIF IN ANODYSED ALUMINIUM BY MAURICE LAMBERT.
MAIN DECK—R.M.S. "QUEEN MARY".

BREAKFAST

California Figs in Syrup   Compôte of Rhubarb   Baked Apples
Compôte of Figs   Compôte of Prunes
Grape Fruit   Apples   Oranges   Bananas   Pears
Orange Juice   Prune Juice   Tomato Juice

Cream of Wheat   Quaker Oats   Oatmeal   Bran Flakes
Rolled Oats   All Bran   Bonny Boy Toasted Oats   Post Toasties
Force Whole-Wheat Flakes   Shredded Wheat   Grape Nuts

Onion Soup Gratinée - (*To order 10 minutes*)

Fried Fillets of Whiting   Grilled Codfish-Parsley Butter
Kippered Herrings   Finnon Haddie in Cream

Eggs:—Boiled, Fried, Turned, Poached and Scrambled
Shirred Eggs and Grated Tongue   Omelettes Various

Hashed Chicken and Mushrooms   Scotch Collops and Poached Egg
Sauté Calf's Liver Sauce Robert

FROM THE GRILL (*To order*)
Ham Steaks-Devilled Sauce   Lambs Kidneys on Toast
Palethorpe Sausage   Tomatoes
Pale and Smoked Wiltshire and Irish Bacon   Wiltshire and York Ham

Potatoes:—Lyonnaise   Mashed   French Fried   Saratoga

ASSORTED COLD MEATS

SALADS   Tomatoes   Watercress   Spring Onions   Radishes

CAKES   Buckwheat and Griddle Cakes   Waffles - Maple Syrup

BREADS VARIOUS   White and Hovis Rolls   Brioches   Crescents
Soda Scones   Triscuits   French Toast   Sultana Buns

Conserve   Honey   Honey in the Comb   Marmalade

Tea:—Indian, Ceylon and China   Coffee   Cocoa
Cadbury's Cup Chocolate   Horlicks Malted Milk - Plain or Chocolate

1936 Breakfast Menu

Queen Mary entering NYC on her maiden voyage

Cabin Class Passengers

Former First- and Third-Class Pool

Observation Bar

Verandah Grill

First Class Library on Promenade Deck

Cabin (First Class Post-War) Class Stateroom

First Class Writing Room on Promenade Deck

Queen's Salon on Promenade Deck

The Midships Bar on Promenade Deck

The Royal Salon on Promenade Deck

Third Class Cabin

Second Class Cabin

Britannia Salon

Third Class Garden Lounge

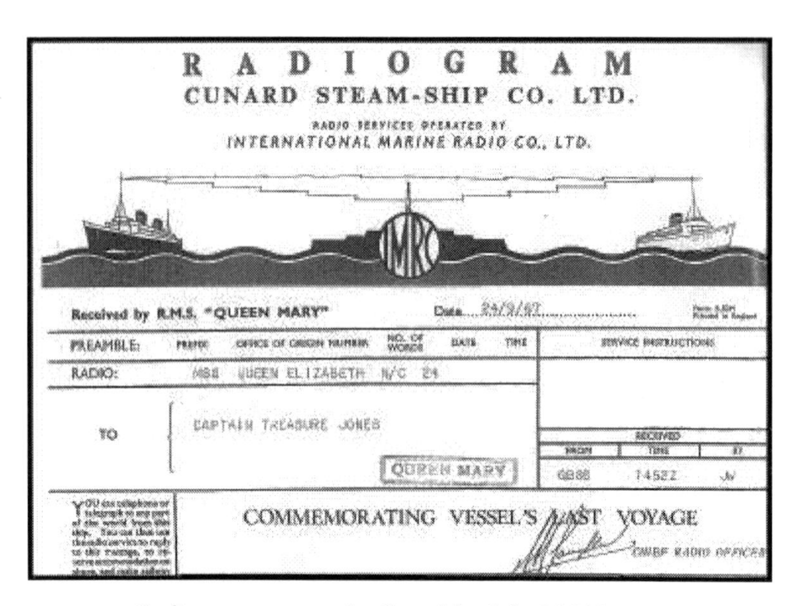

R A D I O G R A M

CUNARD STEAM-SHIP CO. LTD.

RADIO SERVICES OPERATED BY
INTERNATIONAL MARINE RADIO CO., LTD.

IMRC

Received by R.M.S. "QUEEN MARY" Date 24/9/67

PREAMBLE:	PREFIX	OFFICE OF ORIGIN NUMBER	NO. OF WORDS	DATE	TIME
RADIO:	MBB	QUEEN ELIZABETH	N/C 24		

SERVICE INSTRUCTIONS

TO CAPTAIN TREASURE JONES

QUEEN MARY

RECEIVED		
FROM	TIME	BY
GBBS	1452Z	JW

COMMEMORATING VESSEL'S LAST VOYAGE

CHIEF RADIO OFFICER

Radiogram commemorating Queen Mary's final 1,001 voyage

War Brides Aboard the Queen Mary

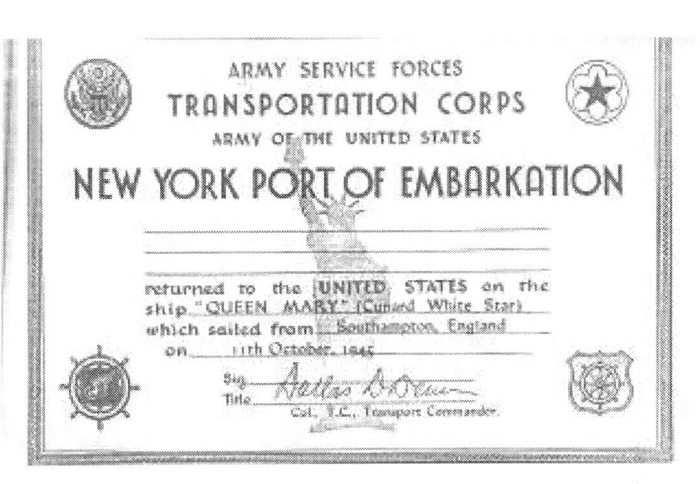
ARMY SERVICE FORCES
TRANSPORTATION CORPS
ARMY OF THE UNITED STATES
NEW YORK PORT OF EMBARKATION

returned to the UNITED STATES on the ship "QUEEN MARY" (Cunard White Star) which sailed from Southampton, England on 11th October, 1945

Sig. [signature]
Title Col., T.C., Transport Commander.

Repatriation Certificate

Final Voyage as a Troopship During WWII

Grand Salon—Cabin Class Dining Hall

First Class Cinema

Drawing Room

WWII Servicemen on Deck

A Deck Second Class Lounge

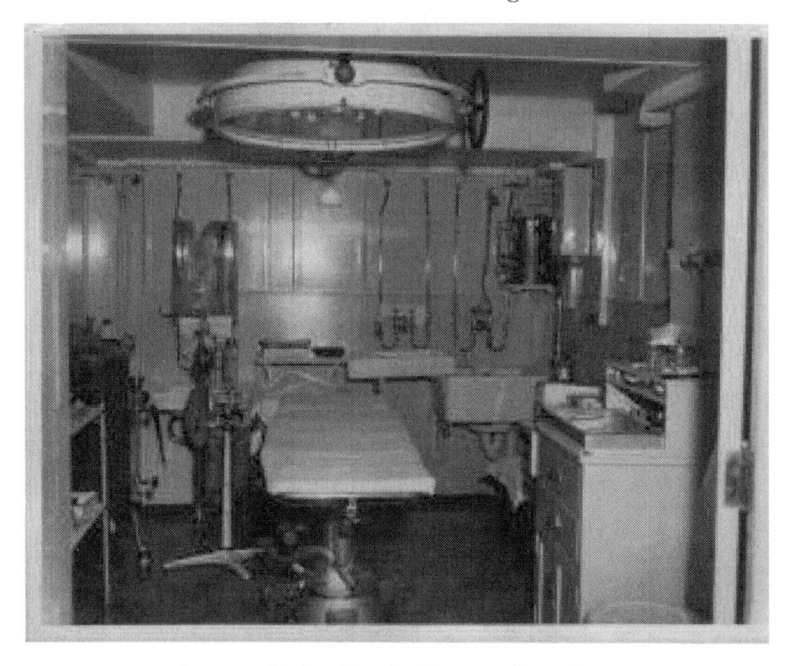

Passenger Services Hospital Operating Room No.2

This photograph was captured on one of the "Paranormal Shipwalk" tours. In the oval, you can clearly see the apparition of a man adorned in a black fedora-like hat and coat. No one physically matching this description was on this particular night's tour. Intriguingly, this ghost, often dubbed "The Fedora Guy," has been seen throughout the Queen Mary. *Photo courtesy of Tony Ashlin.*

Shadow Figure on B Deck

PART 4

Long Beach Welcomes the RMS Queen Mary

"The farther backward you can look, the farther forward you can see."

—Sir Winston Churchill

Photo courtesy of Joe Bertoldo

The *Mary* underwent extensive post-war refitting and refurnishing to enable her to resemble her pre-war luxury years. Some changes were made to the *Mary*, such as a separate movie theatre, new lounges, and a cocktail bar, among others. The shopping center was even improved with better lighting. With a new coat of Cunard color paint, the *Mary* left Southampton in July of 1947 for a small cruise of tests and trials. On July 31, the *Mary* traveled from Southampton to Cherbourg, which marked her first post-war commercial sailing to New York. Not surprisingly, this voyage was completely booked with a total number of 1,957 passengers. Some of these passengers were even on the May 27, 1936 maiden run. For the remaining 20 years, the sister Queens maintained an express run between Southampton and New York.

The year of 1957 marked the best year for Cunard as twelve of its liners were in service. The liners included *Queen Elizabeth, Queen Mary, Mauretania, Caronia, Britannic, Media, Parthia, Scythia, Saxonia, Ivernia, Carinthia,* and *Sylvania.* As jet travel came into the picture, however, life for ship travel and Cunard itself was drastically changing.

THE FINAL VOYAGE

"No finer ship ever sailed this ocean. They will never build another like her. I don't want this to be a nostalgic crossing. We will go out in a blaze of glory and then onto Long Beach. California's climate will be good to her."

—Captain John Treasure Jones

In many ways, the *Queen Mary's* final 1,001st voyage on the seas resembled her very first voyage back in May of 1936. During her inaugural sailing, the world was proud and honored to have such a wonderful ship and people lined by the masses to see her off to the ocean. Thirty-one years later, the world was still proud and honored to have the *Mary* and crowds of people lined up to see her off on her final crossing.

Crew, passengers, and the general public were proud to see the Queen of the Seas off to her element in 1936. Now, 31 years later, their emotions were mixed with pride, joy, and sadness as the ship sailed to Long Beach, California, on her final voyage. In her 31 years, the *Mary* sailed thousands of nautical miles and held numerous fare-paying passengers from all walks of life.

Mayor Lindsay of New York City added, "Some will say that this is a sad day. Rather it is a sentimental day. In some respects it is a wonderful day—wonderful in the sense our city and country are pausing to show this affection and to say hail and farewell to this great ship."

A lavish party was held on the eve of the *Mary's* last voyage, which was titled, "A Farewell Salute to the *Queen Mary*." Crew and passengers celebrated and reminisced about the ship's remarkable 31 years of service. Seven hundred people attended this bon voyage, which benefited the Travelers Aid Society of New York. The Lester Lanin orchestra played music through the night. Passengers enjoyed a gourmet dinner and cocktails in the first class restaurant and danced the night away. On a humorous note, the *Mary* signaled its first S.O.S. in her history, which involved a peculiar petition for help. The *Mary* had seen the last of her champagne bottles.

Honorary patrons for this luxurious party consisted of Sir Basil Smallpeice, Chairman of the Cunard Line; Sir Patrick Dean, British Ambassador to the United States; Lord Caradon, British delegate to the United Nations; Arthur J. Goldberg, United States delegate to the United Nations; and Anthony G. R. Rouse, British Consul General.

The City of Long Beach purchased the *Queen Mary* for $3.45 million from Cunard for her conversion into a hotel and museum. In total, there were 320 offers made from around the world to purchase the ship. Propositions for the *Mary* came from the United States, Australia, Japan, Canada, Hawaii, the Virgin Islands, Gibraltar, Norway, Denmark, Scotland, and other areas of the United Kingdom. Out of the twelve most practical offers, the City of Long Beach's bid was one of the most viable and attractive.

It was suggested that the vessel could house elderly persons or the homeless or even be refitted as a casino. Hugh Hefner even pondered the idea of turning her into a Playboy Bunny ship, which did not go over well with Sir Basil Smallpeice. New York City offered $2.4 million to transform the vessel into a floating high school! All in all, the fate of the *Queen Mary* would eventually fall into the hands of Long Beach, California.

However, at the time of purchasing the ship, it was not realized that the monetary cost for transporting the empty liner from Southampton to Long Beach would be $650,000. Thus, the city along with the Fugazy Travel Service decided it would be worthwhile to make the final journey into a luxury cruise in order to accrue some money for the ship's transport costs.

One-thousand-and-ninety-three passengers paid for the last cruise, which lasted 39 days. Fares ranged from $1,100 to $4,500. The majority of the passengers were senior citizens who hailed from Southern

California. Some had lots of money while others had to save money in order to afford their fare.

The *Mary's* final sailing on the seas began in Southampton and ended in Long Beach at noon on December 9, 1967. She traveled 14,559 miles and sailed for 40 days after departing Southampton. Her ports of call for this last stretch were Valparaiso, Chile, Callao, Peru, Balboa, Panama, and Acapulco, Mexico. Off the California coastline, a Douglas DC-9 tried to re-enact the dropping of roses onto the ship, which was done on her maiden voyage entry into the New York harbor. However, this time, the roses missed the *Mary* and ended up in the sea.

Mr. Horner was one of the *Mary's* veteran crew members. He sailed on the ship's maiden voyage back in 1936. Of course, the liner's last crossing was a sentimental one for him. "This job has been like a home to me. It's like a factory closing down. But when one door closes, another opens up," said Horner.

In December of 1967, the City of Long Beach made history by welcoming the RMS *Queen Mary*, the most lavish and historically significant vessel of all time. As the *Mary* slowly made its way into Long Beach's harbor, it was surrounded by a flotilla of boats and crowds of spectators. In fact, 5,000 vessels awaited the *Mary* as she regally made her way into Long Beach. Cabin cruisers and Coast Guard cutters greeted the ship. Tugs guided the liner to Pier E. At 12:07 p.m., Captain John Treasure Jones signaled "finished with engines" for the final time, directly from the bridge telegraph.

Two days later, he handed over the *Mary* to the City of Long Beach, California. Her propellers were detached from the engines and the liner was eventually classified as a building. Her reigning days traveling the seas came to an end; however, she did not merely retire. In Long Beach, she has continued to share her history and unrivaled success with all of her visitors.

On May 17, 1968, the *Queen Mary* was officially stationed at Pier E in Long Beach harbor after spending over a month in dry-dock as the Navy performed a routine physical on her. She received a fresh coat of paint on her white waterline stripe and underwent work on her outer hull. This was the first and one of many conversions and refurbishments done to change the ship into a hotel, museum, and convention center. Despite her unmatched 31-year career, she appeared to be in relatively good condition.

THE RMS QUEEN MARY'S REIGN IN LONG BEACH, CALIFORNIA

Long Beach is the epitome of innovation and has received numerous awards. Its amenities are world-class, and besides the illustrious *Queen Mary*, it houses the Aquarium of the Pacific, Long Beach Airport, Port of Long Beach, various theaters, museums, marinas, delectable eateries, and a whole plethora of other features for people of all ages. The *Queen Mary* is also a significant economic frontrunner for Long Beach residents and assists in producing tax revenues that fund citywide amenities. In fact, according to Beacon Economics in its economic impact study, the ship has a tremendous influence on Long Beach's economy, such as generating $93.7 million in fiscal output.

The *Queen Mary's* 100,000 square foot museum was located on the lower six decks and featured 40 exhibits. The total cost to build this museum was $14 million dollars. The City of Long Beach itself donated $8.75 million dollars for the construction of the museum, which took approximately four years to complete. Its conversion process involved very tedious and meticulous work to prevent the vessel from receiving any injurious damage. The City of Long Beach decided to permanently move the liner to the 311-acre Pier J.

In 1992, the City of Long Beach shockingly made the decision to sell the liner, having received 18 proposals to purchase the ship. Initially, the city planned on accepting the highest bid of $20 million from Hong Kong. However, it had a change of heart and signed a five-year contract with a non-profit foundation to manage the vessel in Long Beach. The Harbor Commission of the Port of Long Beach gave the ship to the city and contributed toward its restoration.

Spirited Queen Mary: Her Haunted Legend further elaborates that at the time the *Queen Mary* was put up for sale, the City of Long Beach had plans to improve its economy and waterfront. Thus, proposals were underway to build a marina and retail facility, with the hopes of attracting tourists. City officials believed that the idyllic attraction would be a museum devoted to the sea, with the *Queen Mary* being its top-notch setting, serving as a hotel, banquet, and convention site. The Diner's Club Credit Card Company invested $4.5 million in the construction of a luxury hotel, nightclub, restaurant, and first-class shopping facility. However, in the summer of 1970, after spending $5.6 million, the company decided to back out of the project.

On April 1, 1971, Specialty Restaurants Corporation became the leaseholder for all areas of the ship, except her hotel. PSA Hotels, Inc. became the operator of the liner's lodge, with the intention of spending $4 million to change existing cabins into a hotel. The *Queen Mary* tour opened on weekends in May 1971. The tour was then opened daily in June; by the following December, almost a million people had visited the grand lady's decks. The first 150 rooms were made available for lodging to the public in November 1972. The remainder of the cabins were planned to open in the subsequent year. Then, in February 1974, the Hyatt Corporation commenced its responsibility for running the ship's hotel.

As also noted in *Spirited Queen Mary: Her Haunted Legend*, Jacques-Yves Cousteau, the noted oceanographer, was made chief designer and planner for the ship's proposed Museum of the Sea. However, he hired his son to develop new design models. Both father and son then established the Living Sea Corporation, with museum goals to educate the public about marine life and ecology. On December 11, 1971, the Living Sea Museum, said to have the world's largest collection of marine exhibits, opened to great fanfare. It was reported that 4,000 individuals walked through the attraction on its opening day.

Economic hardship during 1973 and 1974 cut into the *Queen Mary's* profits; however, the ship proved its Long Beach success in the years to come. On October 1, 1976, *Queen Mary* Tours, Inc. was developed to take over operations of the Living Sea Museum and historical displays. Proposals were developed to invigorate public attendance with a strategic marketing plan, brand new exhibits, and live entertainment.

On September 1, 1980, the Wrather Corporation assumed responsibility for operating the *Queen Mary*. The company began a $10 million renovation initiative to upgrade and restore many areas of the ship, with the intention of highlighting its 1930s splendor. The corporation closed the Living Sea Museum and exhibits to make way for a 50,000 square-foot Exhibition Hall, specifically for conventions and trade shows. Additionally, Howard Hughes' noted flying boat, the *Spruce Goose*, was to be relocated adjacent to the *Queen Mary* as another tourist attraction. In its new home in Long Beach, the *Spruce Goose* was eventually opened to the public on May 14, 1983.

In September 1987, the Wrather Corporation sold its stock to the Walt Disney Company, which then became the leaseholder and operator of the *Queen Mary* and *Spruce Goos*e venues. In 1992, both the *Queen Mary's* hotel and attractions were closed as the search for a new operator went underway. As mentioned, on February 5, 1993, the Long Beach City Council awarded a five-year lease to the brand-new RMS Foundation, Inc. under the auspices of Joseph Prevratil. Therefore, the *Queen Mary* reopened its doors to the public in late February 1993. This was a great year for the ship as she proudly became listed in the National Register of Historic Places, an honorable distinction for her unprecedented years on the seas.

The RMS Foundation worked closely with the Queen's Seaport Development, Inc., which has a 66-year lease from the City of Long Beach to continue operations of the *Queen Mary's* hotel, attractions, entertainment, banquet facilities, and 45-acre development adjacent to the legendary liner. In 2011, Newport Beach, California-based Evolution Hospitality, LLC, became the management company for the *Mary*. Its responsibilities include the supervision of all operational aspects, including the 314 staterooms, meeting spaces, attraction ticket sales, special events, retail outlets, and onboard restaurants.

In regard to managing the *Queen Mary*, Evolution Hospitality's President John Murphy has said, "Our primary goal is to increase revenues in order to preserve this wonderful community tourism asset and international icon as well as help it flourish. By implementing our sales-driven culture and knowledge of the Southern California marketplace, our team will focus on growing the *Queen Mary's* customer base and leveraging the considerable investments made in the ship's facilities over the past several years."

Further plans are underway to develop a 65,000-square-foot maritime museum and learning center aboard the *Queen Mary*. The Queen Mary Heritage Foundation is eager to announce plans to create this world-class exhibit on the ship. With a strong ambition to educate people on the *Queen Mary's* history, these developing educational programs are designed to enhance children and adults via the arts, sciences, and humanities. The museum will house classrooms and exhibition spaces, a 4D theater, and traveling shows.

Along with many other businesses and attractions, the *Queen Mary* closed its doors during the COVID-19 pandemic. After a long slumber, she awakened and resumed a limited amount of her iconic tours on April 1, 2023. It was a joyous yet surreal feeling to embark with friends on April 15th for the first time since December 2019. You better believe I shed some tears of joy. The *Queen Mary* triumphed once again – surviving the global COVID-19 epidemic.

According to a City of Long Beach Press Release, Urban Commons Queensway LLC, the ship's previous lessee, relinquished its existing leases in June 2021. That same month, the City of Long Beach regained control of the RMS *Queen Mary* for the first time in over 40 years. Its Department of Economic Development manages the famed ocean liner's financial agreements, while the Department of Public Works spearheads structural repairs and preservation endeavors.

In April 2023, the City of Long Beach disclosed a projected partnership between itself and the Port of Long Beach, helping to reinforce the city's tourism, hospitality, transportation, and logistics. The Long Beach Energy Resources would lessen its oil operations in certain areas; thus, affording more land to be developed. This agreement would permit the Port to advance $12 million to the City of Long Beach. This money will be utilized to finance the ship's re-opening, support the ongoing restoration, and pay attention to infrastructure necessities – all important to assisting the *Queen Mary's* revenue streams for future upgrades and development of the adjacent acreage.

Along with previous studies and the assistance of marine engineering specialists, the city proposed the specifications and plans for the mandatory repairs. The anticipated costs for this work amounted to around $5 million. For the repairs to start in February 2022, the State Lands Commission approved its use of Tidelands funding in December 2021 as required by state law. The Long Beach City Council accepted an initial funding cost of $2.5 million.

With the help of Exbon Development, Inc., restoration included the removal of decrepit lifeboats, which were applying stress on the side shell of the vessel, generating substantial cracks in the support system. The *Queen Mary's* structural stability was improved upon their removal. Two lifeboats remain for future preservation efforts.

Other restorative undertakings included HVAC repair, the installation of new permanent bilge pumps, new boilers and heat exchangers, a water intrusion warning system, improvements to the bulkhead, Promenade Deck gangways, and elevators, among others. Refining the surrounding circuits, electrical timeclock, and lighting fixtures safeguards the electrical systems and appropriate usage of lighting.

Ongoing restorative efforts are ensuring the *Queen Mary's* longevity for current and future generations. Her best years are yet to come.

Today, the *Queen Mary* is a magnet for tourists from around the world. She is a legend and has stirred the imagination of folks from all walks of life. It has been said that approximately 1,000,000 people a year visit the *Mary* to learn about her history and join in her transatlantic splendor. In addition to her museum, she is equipped with stunning views, award-winning restaurants, shops, and seasonal attractions. Travelers can stay in one of the 314 hotel staterooms, which have been preserved to resemble their original décor. Additionally, the ship also offers a variety of tours and hosts special events, such as reunions and weddings. When planning your visit to California's iconic southern region, make sure to have the *Queen Mary* at the top of your list.

The Queen Mary's Haunted Voice: Paranormal Occurrences Aboard the Ship

"If we open a quarrel between past and present, we shall find that we have lost the future."

—Sir Winston Churchill

Photo courtesy of Joe Bertoldo

This book would certainly not be complete without giving an overview of the Queen's haunted voice—that is, the paranormal occurrences that exist throughout her inner walls and hull. There are many spirits that reside on the ship; however, there are certain ethereal energies that are more widely known among the ship's employees and visitors. In addition to discussing her well-known energies, this section will include ghostly stories from the author, as well as from some of the author's paranormal colleagues and friends. Additionally, this portion of the book will include some paranormal encounters from various ship tourists, extending from the 1980s to the present. Finally, this section will discuss some common theories for paranormal existence and delve into some hypotheses that might explain why the *Mary's* paranormal activity makes her one of the most haunted spots on this planet. In my book, *Spirited Queen Mary: Her Haunted Legend,* I elaborate more in detail on the liner's various ghosts and spirits as well as further discuss her hot spot locations and their paranormal experiences.

The RMS *Queen Mary* is a global icon. She's widely known for her unprecedented history and well-known paranormal phenomena. As such, she sadly falls into the hands of the entertainment industry, one that often exploits, exaggerates, and sensationalizes her paranormal activity. This has led to a whole cacophony of embellished and erroneous fabrications for her ghostly happenings; thus, further entrusting *Queen Mary* historians and/or those who are quite familiar with the ocean liner's resident spirits with the task of educating the public on the authenticity of her ethereal claims.

There is a lot of inaccurate commentary on the Internet regarding the otherworldly occurrences aboard the *Queen Mary.* I implore you to not believe everything you read or watch on television or the World Wide Web. Please honor the ship's everlasting legacy by applying your intelligent discernment over the ethereal encounters you may have and/or ones you've heard from the grapevine. If you're unsure or need more explanation, please contact a trusted source, such as myself or a *Queen Mary* staff member.

All genuine ghostly events aboard the vessel shape her storied tapestry. They should be treated with respect and reverence as they all help to relay the story of the greatest ocean liner to grace humanity. As we know, there is a mix of residual and intelligent spiritual activity. Her resident energies all have unique stories and they all deserve to have them chronicled with esteemed accuracy. They all contribute to the *Queen Mary's* living legend.

The Ghosts and Spirits of the RMS Queen Mary

J.P.

Out of respect for this individual and his family, I will only use his initials when discussing him. A young man from Skipton, North Yorkshire, J.P. joined the *Queen Mary's* crew in March 1966 as a fireman and bilge cleaner in the #3 boiler rooms. At around 3:55 a.m., on July 10, 1966, J.P. was found crushed and unconscious in watertight door #13 in the shaft alley's starboard side. Sadly, he later passed away in the ship's infirmary at the tender age of 18 years.

I considered two initial theories on the origins of this catastrophic accident when I first embarked on researching the *Queen Mary* some twenty years ago. I wondered if J.P. tripped while attempting to enter watertight door #13, got back up, and couldn't get through in time. Or, perhaps he dropped one of his tools, motioned to pick it up, and misjudged the timing of the closing door. As the years followed, I realized that both of these were less likely to occur based on historical records and ship status at the time of his passing. As my knowledge of the *Queen Mary's* design and history further emerged, I settled on one reasonable yet instinctive theory on what happened that fateful July morning of 1966.

It's important to mention the SOLAS maritime laws, which were initiated after the RMS *Titanic* tragedy on April 15, 1912. The Safety of Life at Sea Act authorized the closing of all lower area watertight doors as a way to prepare for a potential imminent collision with a target unable to be seen due to poor visibility. While out at sea, shortly before J.P.'s tragic demise, the *Queen Mary* was met with an intense fog layer that enveloped the sea. As a result, wheelhouse crew members followed maritime regulations and ensured that all boiler and engine room watertight doors were closed on bridge control and in good working order. The watertight door light panel in the wheelhouse indicated that #13 had either a mechanical malfunction or interference; thus, calling for an immediate examination. Much to their dismay, fellow engine workers found J.P. crushed in between the hydraulic door frame and panel.

Openings in the watertight bulkheads are equipped with 38 power-controlled sliding watertight doors and 28 hinged-type watertight doors. Deptford, London's Messrs. J. Stone & Co., Ltd., developed the hydraulic power system responsible for managing their opening and closing. A panel on the navigating bridge indicates the condition of each door; each one contains a colored disc and engraved number. The discs lit up and continued their illumination while the doors remained closed. A bell was located at each of the doors and sounded approximately seven seconds before closing commenced. Crew members were able to independently operate the doors by hand against the pressure system by levers located at either side; thus, safeguarding people from being trapped in a compartment upon the bridge's shutting of the doors.

To our knowledge, no one witnessed what happened to J.P., but the fascination with how such a tragedy occurred on a ship with such revered credentials continues to prevail. Although various theories abound, I've felt strongly about one in particular: in my opinion, I believe J.P. was trying to maneuver the manual hydraulic release levers to open the door when he misjudged the timing of it and couldn't get through. Since the doors were shut as the ship entered the fog, it makes logical sense that J.P. was attempting to open watertight door #13.

As with others, my research into this incident has been a combination of archival research, interviews with former engine crew members and other engineering specialists as well as hours spent studying the aft engine room. As mentioned, I logically suppose that J.P. endeavored to open watertight door #13 and misjudged the short time it took to completely shut. It took a mere six seconds for it to close the maximum of three feet. Let's suppose J.P. opened the door ajar and tried to slide through, not realizing that he only had a second or two to do it. Or, perhaps, he didn't realize that watertight door #13 would automatically start shutting. The power system was designed to close the doors automatically after they've been manually opened.

One summer evening in 2017, I was a guest on the "Paranormal Shipwalk" tour hosted by my friend Tony Ashlin. Tony relayed to the group that a former aft engine crew member was aboard the *Queen Mary* during J.P.'s last voyage. While he attended Mr. Ashlin's tour one evening, he shared the same aforementioned opinion on what led to the untimely tragedy. He, too, felt that J.P. was attempting to manually open watertight door #13 before trying to get through.

It's quite apparent that J.P. felt pride and joy as a fellow crew member of the RMS *Queen Mary*. To this day, he remains one of the famed ocean liner's most likable resident spirits. I conversed with a former *Queen Mary* waiter in 2019 who shared a neat story of how he met J.P. the night before his tragic passing. While reading his account, I sensed J.P.'s affable nature and wonder at the vessel's magnificence. Although I promised to not disclose the details of the rendezvous between these two men, I will say that J.P. seemed happy, content, and overjoyed as an esteemed crew member.

Please let sound judgment and reason reign over the excessive, sensational, and exploitative myths surrounding J.P.'s heartbreaking demise. Do not even engage the various histrionic and implausible presumptions as to what occurred on that fateful morning of July 10, 1966. This young man's passing should not be used as a stage for individuals to promote offensive and theatrical claims associated with his accident. My home-driven point here serves as a reminder to humanize J.P. and honor his short yet dedicated work aboard a ship beloved by so many. Honor the memory of this young lad who set off on a journey to work aboard the greatest ocean liner ever built and in return, became an everlasting part of its legacy.

Below is the actual *Queen Mary* (courtesy *Queen Mary Archives)* archived information regarding J.P.'s death:

10 Jul—07:00 REPORT OF CREW DEATH, 10th July 1966, AT SEA, RMS QUEEN MARY:

At 7:00 a.m., on Sunday 10th July 1966, I was informed by Mr. Burgess, Senior Second Officer, I/C 4-8 watch that an accident had occurred during his watch at 3:55 am (clocks retarded one hour), when J. P. F/Cleaner R832490 had been found trapped by No. 13 watertight door in starboard tunnel, receiving severe injuries from which he died. At the time of the accident, the vessel was in fog and all watertight doors in boiler and engine rooms were closed on bridge control. When examined by me at 7:15 a.m., the fog had cleared, doors were on open position and on testing No. 13 watertight door (horizontal type) was found to be in good working order taking 6 seconds to close the maximum of 3 feet. There was no oil on the plates in the vicinity of the door and J. P. had been engaged for three voyages as F/Cleaner in No. 3 boiler rooms and was making his second voyage on bilge pumping duties in engine rooms and tunnels.

THE LADY IN WHITE

There have been numerous sightings of a ghostly lady in white aboard the *Mary*. This ghost is commonly seen in the hotel lobby bar area by the piano. She has also been seen in the Observation Bar and Queen's Salon. Witnesses describe her as wearing a white evening gown and she is often seen dancing by herself.

Many hotel guests will even hear the lobby bar piano playing by itself when they arrive at the front desk. Some people feel that this ethereal lady is somehow connected to the popular instrument. Others believe that she may have been one of the ship's nurses.

I have another theory as to the lady in white's origin. One of the employees witnessed a female apparition adorned in a white gown travel out of cabin 444 and glided forward toward the Isolation Ward area. The number "444" is often considered an angelic number. Thus, I wonder if the lady in white is an angelic being who comes aboard the ship to provide peace and serenity.

WINSTON CHURCHILL

The ghost of Winston Churchill has been spotted many times on the *Mary*. It has been said that he likes to frequent the decks while smoking his cigars. Recently, his apparition has been spotted looking out toward the Long Beach harbor while on the sports deck areas. Many people, including myself, have smelled phantom cigar smoke by the Winston Churchill Suite on M Deck. Many people have also heard disembodied voices emanating from the Churchill Suite as well. He is also known to walk around the gift stores on Promenade Deck.

One of the house-keeping staff members had an interesting encounter with Churchill's spirit as she entered his former suite on M Deck. She relayed that when she opened the door to his suite, she saw a man sitting in the center of the room. When she entered, the man looked at her, turned and floated into a portrait on the wall. The portrait was that of Winston Churchill!

Allen from Las Vegas, Nevada, recently stayed in the Churchill Suite aboard the *Queen Mary*. In his own words, he describes the following:

> *We stayed in the Churchill Suite for four nights. Great room and we heard it was part of the ghost tour. Nothing happed the entire time; we didn't experience any smell of cigars or anything else moving, etc. However, the last night (our wedding night), we were very tired after a long ceremony, pictures, and dinner. We were in bed, and exactly at 1:00 am, we heard the closet door open and wire hangers moving around and then the door close. I shot up along with my wife Cierra, and I said "what the f***" and for some reason looked at the clock and it was exactly 1:00 am! We sat there in shock for a bit and I walked around and nothing. The next morning, we checked every closet and we had no wire hangers in any of the closets. That room is the quietest suite on the ship. You can't hear anything in that room, but the closet doors and hangers seemed to be right in front of us and none of the doors moved. Not sure what happened but it was strange.*

JACKIE

When you travel to the *Queen Mary,* you will hear of the various stories regarding the famed spirit Jackie. Some people have surmised that Jackie was the child of a crew member or a child who drowned in the second class pool. However, there are no records of a child by the name of Jackie passing away in this area or any area of the ship for that matter.

Jackie seems to be around five or six years of age and is most commonly seen, heard, or felt in the former first- and third-class pool room. She is very loquacious and has been witnessed singing and laughing. Jackie has also been seen wandering the halls of A Deck as well. All in all, she can be experienced anywhere on the ship. Even though Jackie is talkative to many people, she is drawn to certain individuals.

In my many years of conducting paranormal research aboard the ship, I have developed a strong rapport with Jackie, as she is very intelligent and a highly evolved spiritual energy. I have had the wonderful opportunity of seeing her twice as well as hearing and capturing her voice via audio recorders. I do believe that she is, in a way, the spiritual ambassador for the ship. I give a more thorough account of Jackie in my book, *Spirited Queen Mary: Her Haunted Legend,* as well as share some of my most profound experiences with her.

As we know, increasing theories exist attempting to explain the origins of Jackie. However, there's another one that I've pondered over the years that may shed light on the genesis of this beloved resident spirit. It's a personal presumption shrouded in the same enigmatic mysteriousness often surrounding paranormal phenomena. Is it possible that Jackie's mannerisms and personality are the embodiment of the ship itself? In other words, is Jackie's ethereal existence the personification of the *Queen Mary?* The uncanny similarities between the two are almost undeniable: femininity, innocence, endurance, sentience, and wonder. Perhaps, the spirit of the celebrated *Queen Mary* is portrayed through the sightings and encounters people have with Jackie.

Based on my twenty years of visiting and researching this illustrious ocean liner, I've nearly concluded that there's a lot more to the real Jackie besides just being an earthbound entity often experienced by crew and visitors alike. There's a deeper meaning to her transcending the conventional thoughts of who she is, one that may not be meant for mortals to truly understand.

It's quite intriguing how the name "Jackie" originates from English roots and means "God is gracious." Our ultimate higher-up is very gracious for granting the world the true gift of the RMS *Queen Mary*. My mission prevails to uncover Jackie's authentic beginnings and I've already surrendered to the thought that I may not entirely know until it's my turn to enter the gates of the afterlife. Until then, I will continue searching.

SARAH

Many individuals with intuitive abilities have sensed a child by the name of Sarah in the former first- and third-class pool area, specifically in the confines of the dressing rooms. One of my colleagues captured an EVP in the dressing stalls of a young woman answering "Sarah" when asked who the investigator was communicating with. I also captured an EVP disclosing the name of "Sarah" in a stateroom on B Deck. Since the name is fairly common, my capture could have referred to someone else. Is Sarah a made-up haunt, a psychokinetic creation if you will, due to the amount of energy projecting her name onto the changing rooms?

Those claiming to have witnessed the entity allegedly named Sarah describe her as being around eight years old. Some people even feel that she is in her teens. She is depicted as being thin and lanky with short dark brown hair. Some have described her energy as being gruff and forceful; however, I have not personally sensed or encountered this. As with Jackie, no one is exactly sure of her origins or whether she has a history with the *Queen Mary*. There is speculation that she may have been one of the third-class passengers. However, there is no forensic evidence of a documented death of any child named Sarah on the ship.

YOUNG WOMAN IN THE POOL ROOM

The former first- and third-class pool room is one of the *Queen Mary's* most haunted areas. There are various ethereal energies residing in the pool area. One such ghost is that of a young woman who is seen wearing a miniskirt. Witnesses have described seeing a young woman walking down the stairs and then disappearing or even jumping into the now empty pool. She has also been seen peeking around the columns that support the second story of the pool room. No one really knows who this ghost is or how she died.

OTHER SPIRITS IN VINTAGE BATHING SUITS

The former first- and third-class pool is certainly not complete without seeing phantasms adorned in vintage bathing suits as they take a ghostly swim. Many visitors have witnessed the apparitions of swimmers splashing around in the water. One individual even relayed that he saw the pool filled with water, even though it has been waterless since the late 1980s.

Quartzite, a strong retainer of spiritual energy, can be seen on the pool room's ceiling. Furthermore, this area is considered the heart of the ship, where passengers visited daily to relax and exercise. Think about all of the emotions exchanged between people in this area! Thus, there must be some sort of imprint containing these sentiments. The pool is one of the most visited areas on the liner and scores of people tour the area daily. There may be some sort of psychokinetic interplay responsible for its numerous paranormal sightings. In addition to the feelings of ghosts and spirits, emotions from living individuals may be retained in the confines of this particular location.

WILLIAM ERIC STARK

William Eric Stark was one of the *Queen Mary's* Senior Second Officers. Unfortunately, he died a tragic death while aboard the ship. Apparently, the staff captain told Stark to invite the two watch officers to have some gin. A steward was asked to acquire the alcohol and located a bottle of what he thought contained gin. However, that particular bottle contained the deadly liquid that killed Stark. Stark mentioned to the staff captain that he had consumed tetrachloride and lime juice, obviously confusing it with gin. At that time, Stark was not aware that his demise would be quickly approaching.

The following day, Stark started to physically feel the consequences of downing tetrachloride. The doctor on duty advised him to rest. On Tuesday, it was recommended that Stark be transferred to the hospital as he deteriorated rapidly. He eventually fell into a coma and later died.

The ghostly officer is seen re-enacting his final moments on earth. He is seen walking along the deck as he would do during his duties. He has also frequented the third class areas. In fact, since he was known to smoke cigars, some ship guests have smelled the strong phantom aroma of cigar smoke in these areas.

Since Stark died a horrible death aboard the ship, it is no wonder why his ghost remains. Unfortunately, he may not be at rest, as he may not know he is dead and continues to go about his work-related responsibilities. If you are visiting the *Queen Mary* and are offered a drink of gin by a ghost, you will most likely know who it is. It is my sincere hope that Mr. Stark, as well as all other earth-bound energies, find peace and solace.

KITCHEN GHOSTS

Yes, the *Queen Mary's* kitchen is even said to be haunted. There are various stories as to why the kitchen is said to house ghostly phenomena. One story relays the possibility that a former cook was reportedly pushed into a hot oven because he apparently had a negative attitude. Another variation of the same story suggests that the cook was shoved into the oven when a fight broke out among American troops. These are just rumored accounts and have not been verified by fact. People have heard, seen, and felt the presence of the ghostly cook while inside the location. His demeanor has been described as being angry and irritable.

A totally different story might suggest why the kitchen continues to cook up ghostly events. Leonard "Lobster" Horsborough was a former *Queen Mary* cook who worked on the ship for 15 years. He died on November 13, 1967, from a combination of heat stroke and cardiac failure. Captain Treasure Jones buried him at sea among a crowd of people on A Deck. Horsborough loved working on the *Queen Mary* and did so for many years. Therefore, his spirit may choose to reside on the ship he loved.

BOILER ROOM SPECTERS

The former boiler rooms also have paranormal activity from time to time. The residual screams from dying men and the sounds of rushing water can be heard in the exact spot where the *Queen Mary* penetrated the HMS *Curacoa.* This phenomenon can also be experienced in R Deck Forward. Additionally, the ghosts of soldiers can be seen looking down from the red bridge that is used as a passageway during paranormal tours. The phantom sounds of men working can still be heard in the former boiler rooms as well. The Green Room is an area that was formally used by Disney cast members. Today, it's an empty room that's full of concentrated spiritual energy. Shadow figures, disembodied voices, and numerous electronic voice phenomena captures have occurred in this small area. *Spirited Queen Mary: Her Haunted Legend* gives a more detailed account of the paranormal energy in the boiler rooms.

JOHN HENRY

Many intuitive individuals feel that a boiler room worker passed away from a tragic accident while on his shift. Interestingly, a few psychics claim that this individual succumbed from scalding burns while working in the boiler rooms. Furthermore, this male spirit seems to consistently respond to investigators when they mention the name, "John Henry."

Obviously, there is no forensic evidence to pinpoint whether this incident actually occurred. With that said, many people have spotted a tall, bald male apparition, dressed in vintage work attire, walking around in boiler room #3 and #4. John Henry has also communicated with people in the Green Room.

A Spirit Possibly Named Amanda

For years, I have sensed a female spirit in the boiler rooms, specifically in the #3 and #4 areas. Furthermore, the ship's paranormal ambassador and a former *Queen Mary* tour guide also have experiences with this female energy. This spirit form has always been quite strong: a young woman with a prevailing forlorn and sad existence. I have heard a female weeping in this area as well. When the Paranormal Investigation tour guide shared his encounters with this particular spirit, it did not take too long to realize that we were talking about the same entity. This woman is said to frequent the former first- and third-class pool area as well. In fact, an EVP of a female ethereal entity identified herself as "Amanda" in this exact area. The apparition that has been witnessed in the boiler rooms matches almost identically to one of the entities seen in the pool area. I have always wondered who this woman is and why she makes the *Queen Mary* her ethereal home.

Shadow Energies

The upper levels of the former first- and third-class pool are where most people see the majority of humanoid shadow figure sightings. Within the past few years, this phenomenon has occurred consistently in these two areas. On numerous occasions, I have witnessed both partially and fully-manifested shadow forms rapidly darting forward or aft on these upper heights. Thousand Oaks resident Rachel Ashleman was visiting the *Queen Mary* when she snapped one of the most memorable photos of her life. In this picture, you can see what looks like a human-shaped shadow figure either ascending or descending the forward staircase.

Shadow Figure in Former First- and Third-Class Pool—Courtesy of Rachel Ashleman

As with the former first- and third-class pool room, the boiler rooms are also notorious for displaying shadow figures darting about the area. I have interviewed many people who have reported seeing both apparitional sightings and shadow entities in the third and fourth boiler rooms. Not too many people have investigated the first and second boilers; however, one of the special effects tours does migrate through those locations.

There is one spot in particular that supplies a good vantage point for seeing these ethereal beings. When you are positioned on the Disney stage in the fourth boiler room, look forward and you will see various sightings on both the port and starboard sides. You will also notice a walkway which leads to a black iron gate. This gate leads you into the first and second boilers, with the Green Room just past the gate's entrance on the right. This exact area is a hot spot for paranormal activity. Many people, including myself, have not only seen ghostly sightings but have also heard disembodied men's voices and screams and have further captured audio evidence of the spirit world.

Additionally, pay attention to the catwalks and upper areas of the boiler rooms. People have reported seeing shadow energies and partially manifested apparitions there. A good example of this is evident in the aforementioned Green Room, where visitors have seen a ghostly male's face peering down at them through the ceiling's opening. We think that this could be John Henry. Furthermore, I have seen peripheral views of vintage-clad men walking up and down ladders and moving about the catwalks.

THE MAN IN THE DRESSING STALLS

Another well-known and territorial male energy exists inside the confines of the former first- and third-class pool's changing rooms. Many investigators have captured audio evidence of a man toward the portside area of the stalls near the old toilet room. Personally, I feel that this male is a former crew member who seems to be still making his rounds on the ship. Some people feel that this energy is the spirit of William Eric Stark, one of the crew members who passed away while working on the ship. There will be more information on Stark in the hotel deck chapter.

Some intuitive individuals believe that this man is upset about monetary issues. In fact, on a research project, one investigator captured an EVP of "Yes" after she asked the question, "Sir, are you upset about money issues?" As said before, I do believe that this entity is somehow reliving his duties on the ship, possibly not realizing he is deceased. It is rather sad to experience earth-bound energies, as my ultimate wish for them is to find peace.

GHOST CAT(S) IN DRESSING STALLS

The dressing room of the former first- and third-class pool area is home to another spirit of the furry kind: a cat! Back in the ship's sailing days, it was not uncommon to have livestock aboard the ship. This extends to cats and dogs. Who would think that the spirit of a kitty would reside in the dressing stall area? Let me tell you, folks, there is most certainly a talkative and cute little phantom kitty wandering these halls. He or she has been heard and even seen walking on all four legs. It is said that this particular spirit cat, while alive, belonged to a former crew member. Lately, many people, including myself, have had experiences with this furry specter. I share a spine-tingling account with one of the *Queen Mary's* furry friends in the book *Spirited Queen Mary: Her Haunted Legend.*

PRISONERS OF WAR (POWS)

As briefly discussed, many sensitive individuals have felt the presence of an Italian POW in the former first- and third-class pool area. During the ship's heyday as a World War II troopship carrier, many German and Italian POWs were held captive on the liner. The forward cargo hold is one such area where these prisoners were housed. Apparently, there was standing room only in this location and meals, consisting of bread and water, were lowered down in buckets. As one can imagine, many of these men passed away while on the ship and their physical bodies were unceremoniously thrown overboard while the liner was at sea. This is a tragic reminder of the brutality that occurred during World War II.

THE GUYS UNDER THE STAIRS

Many people have claimed to have odd feelings under the staircase in the pool area. Visitors have said that they feel unwelcomed in this confined area, having the sense that they're being watched by disembodied and territorial eyes. One or two male ethereal energies seem to make this claustrophobic spot their home.

I have explored this site a few times only to feel as if I am not supposed to be present. Some investigators feel that the energy is unpleasant, as if being around a person in a bad mood. Perhaps the entities in this location just want to be left alone, similar to the male spirit in the dressing rooms. If people were ill-tempered in life, chances are that they will be in death, especially if they remain as earth-bound entities.

FORMER SHIP COMMODORES AND CAPTAINS

Whether they appear in residual and/or intelligent spirit form, some of the ship's former captains have made themselves known to visitors. Both Commodore Britten and Captain Treasure Jones have been visually encountered in various areas of the ship. I have seen Captain Treasure Jones in the after engine room, which is, to date, one of the most compelling apparitional sightings I have ever had. Just recently, a guest in the liner's wheelhouse encountered the apparition of a captain in the forward portside area near the telegraphs. A guest also reported seeing a commodore saluting the propeller.

Keep in mind that the liner's former crew did not wear name tags. Thus, if you were to see someone adorned in vintage crew attire without wearing a name badge, then perhaps, you have made contact with one of the ship's ethereal crew members. A good idea is to study the various old-fashioned looks of maritime uniforms.

The ship is known for having re-occurring and continuous types of paranormal activity. Almost on a daily basis, tourists report the phantom sounds of knocking and tapping on stateroom doors and walls. I have encountered this many times. Doors will mysteriously open up or unlock by themselves. Throughout many areas of the ship, phantom footsteps can be heard walking right past you. Many people have reported hearing banging and hammering throughout the ship. Lights will turn on and off and disembodied voices can be heard almost anywhere. Cold spots and the feeling of rushing air are felt even in air-tight areas with no vents or logical drafts. Tourists' belongings have been reported to go missing or moved to another location. Again, in my book, *Spirited Queen Mary: Her Haunted Legend,* I delve more in detail as to the various ghostly encounters aboard the ship.

THE BOY IN NAVY BLUE

Some guests have described a toddler-aged boy in and around the ship appearing in Navy Blue clothing and/or with a blue light surrounding him. He has been spotted in various staterooms and other areas of the liner, including Promenade Deck. Some investigators feel that his name is either "Daniel" or "William." I came into contact with this little spirit in the former first and third class pool room. While there, I saw a small male child with curly, reddish hair adorned in what looked like a miniature sailor's outfit appear and disappear right before my eyes. Danny Rangel came into contact with this little spirit while he was conducting one of his tours. Near the front entrance to the pool, he saw this little ghost in his peripheral vision dressed in a blue sailor's suit. Since this entity's origins are not known, it leaves us with similar questions that we have proposed regarding the genesis of Jackie.

GHOSTS OF WWII SERVICEMEN

The spirited sightings of World War II soldiers have occurred throughout the *Mary*. Many of these encounters are a residual, psychic imprint of times gone by that remain within the ship's confines. Many visitors have seen apparitions of servicemen marching along the hotel deck hallways and the upper levels of the former first and third class pool. However, people have spotted these ghostly forms in various areas of the liner.

LIGHT ANOMALIES

As with other areas of the ship, the former first and third class pool area and women's changing rooms are known for weird light anomalies and orbs. There is a lot of controversy over the orb phenomenon, which has ultimately led to a great orb debate among researchers. Honestly, most orbs can be naturally explained due to insects, dust, moisture particles, lens refractions and flares. With this said, I do believe that some light anomalies may have a spiritual explanation. People have witnessed intelligent moving orbs aboard the *Queen Mary* in various hot spot areas. Some of the criteria that I look for when examining orbs for potential spiritual origins are as follows:

a) Does the orb self-illuminate or is it illuminated by a natural light source?
b) Does the orb show intelligent movement or does it follow the natural direction of air-conditioning currents or air being blown in from the window?
c) Does the orb appear in 3D mode or in front of an object?
d) Does the orb have a distinct trail of light?
e) Does it have highly distinguishable facial appearances or numbers?
f) Does it appear in other colors that could not be caused by a reflection?
g) Are there orbs in the frame or do they appear by themselves?
h) Do orbs appear in photos taken in dusty or moist conditions?

QUEEN MARY PARANORMAL ENCOUNTERS IN VARIOUS SHIP LOCATIONS

I have visited the RMS *Queen Mary* on numerous occasions, and I have had the pleasure of experiencing something supernatural each time. It does not really matter as to what time of day it is, as the ghosts and spirits aboard this ship work around the clock. There are times when the ship seems to be more active, specifically relating to important historical dates or anniversaries. I have noticed that paranormal activity increases when the ship is more crowded with events and tourists. Even though I have experienced paranormal phenomena throughout the *Mary*, most of my experiences have been located in the so-called "hot spots." The following is just a sampling of some of my experience and interactions with the spirits of the *Queen Mary*.

HOTEL DECK HALLWAYS

There is a location on the *Queen Mary* that seems to be paranormally active on a continuous basis. This area is situated at the forward section of B Deck where the stairs are located, leading down to R Deck. Intuitives have identified this area as a passageway for spiritual activity.

During many of my visits to the *Mary*, I always schedule ample time to sit in this general area to see if I can acquire and/or experience anything out of the ordinary. I usually sit down on the stairs for awhile and get a feeling of the environment around me. In this area, I have held what many paranormal researchers refer to as a vigil. I have experienced and captured some incredible evidence in this exact spot.

In this location, I have managed to have a few personal paranormal experiences, which have been backed up by scientific equipment. On one such occasion, I was doing an EVP session, as I sat on the stairs leading down to R Deck. I started to get the sensation of nausea, which lasted for a few minutes. I reported into my digital recorder that I was feeling nauseated. During this time, I asked the question, "How many of you are with us tonight?"

B DECK

After review and playback of this audio session, you can hear a ghostly male's voice answer my question by saying, "Ten." Approximately 20 seconds after I captured this EVP, I documented that I saw a grey mist travel past me on the stairs. This incident left me with a valid question: Can ghosts see each other, as well as the living?

On a different stay aboard the ship, I had another intriguing personal experience while sitting on these same stairs. As I was relaxing on the stairs, I witnessed a fast-moving shadow figure come in through the door on the starboard side of the ship. I could audibly hear this entity's footsteps as it proceeded up the stairs leading to A Deck. I immediately got up and chased after this figure. The hallways aboard the *Mary* are extremely long and there was no way for this entity to have made it to the other side of the ship in the time it took me to travel up to A Deck. Additionally, I did not hear any doors opening or shutting, so I ruled out that any living person entered one of the staterooms. What is interesting is that I remember hearing this figure's footsteps on only part of its ascent to A Deck. Thus, I have surmised that it de-manifested while it was traveling up the stairs.

There is an area on B Deck of the ship, which emanates feelings of despair, sadness, and frustration. I am empathic and every single time that I have traveled through this area of B Deck, I have sensed these feelings. On one occasion, as I was doing an audio session in this area, I saw a grey mist move right past me at the same time that I felt as though

I was being watched. Other investigators and I have both reported disembodied footsteps in this area as well.

Additionally, I have picked up a couple of EVPs in this spot of B Deck, which coincide with the feelings associated with this particular location. On two separate visits to the *Queen Mary*, I have captured a disembodied voice saying "Help me" in this area of B Deck. During World War II, a fire did occur in this vicinity on B Deck; thus, was I tapping into the residual imprints left over from this incident?

On two separate visits to the *Mary*, I have witnessed shadow figures on the northeast starboard side of A Deck. As I was walking down A Deck toward my stateroom, I witnessed a shadow figure move from left to right and disappear into the wall. The second time, I witnessed an entity as I was walking down the same passageway toward my stateroom. This time, I saw what looked like a shadow figure peak outside one of the cabin doors. These particular ghostly sightings are prevalent on the hotel decks.

HOTEL QUEEN MARY STATEROOMS

A167

I have had the pleasure of staying overnight aboard the Queen many times. In February of 2008, a friend and I decided to travel up to the *Queen Mary* and stay for the entire weekend. We stayed in stateroom A167, which is located on the starboard side of A Deck.

On the first night, I had a phenomenal experience in this stateroom. This occurrence was an example of having a personal experience backed up by a scientific instrument. Whenever this happens in the field of paranormal investigating, your evidence is more validated and corroborated.

It was around 2:00 a.m. and I decided to do a mini-EVP session. About ten minutes into my session, I directly saw a grey mist enter through the stateroom's entrance door. I then stated into my recorder, "I just saw some grey mist enter the room, so I know someone's in here with me, so can you please tell me your name?" I ended up getting a great EVP from this question. Upon playback and review of my audio, you distinctly hear the name "Michael" right after I asked the above question.

I had another awesome paranormal encounter while staying in A167. This particular experience occurred the following morning around 11:00 a.m. The night before, I distinctly shut and latched the bathroom door prior to going to sleep. In the morning, I awoke to the sound of someone rustling about in the room. I looked at my friend who was still fast asleep. I then proceeded to put my head back down and continued to stay awake.

Then, I heard the distinct sound of a door knob twisting. I sat up and saw that the bathroom door had been opened by itself! The door was cracked open a little and I managed to see what looked like the bottom portion of a leg and foot in the bathroom. At this point, I put my head back down and rolled over on my left side with eyes wide open.

Then, something incredible happened: I distinctly felt the pressure of something pushing down on me; however, there was no one there. It was like a tingly and prickly sensation, which is how you may feel if a ghost or spirit touches you. I did not feel threatened as this unseen force was touching me. I actually remember feeling peaceful. Thus, I have surmised that maybe this spirit was trying to hug me or make sure I was okay.

M001

In September of 2008, I visited the *Mary* for the weekend with some friends. We stayed in cabin M001, which is located on the starboard side of M Deck adjacent to the suites. In the middle of the night, one of my friends abruptly woke me up claiming that she felt something sit at the foot of her bed. When I woke up, I looked toward the foot of our bed and did not see anything. However, due to my friend's frantic state, it was clear that she did experience some unseen force sit on the bed.

The third person stayed in the servant's quarters, which is adjacent to the main room. The next morning, she came out and told us that she felt something grab her foot several times during the night. She reported to us that she told whoever it was to stop bothering her. Was this the same entity that supposedly sat on our bed in the main room?

Additionally, I took several pictures of this particular stateroom with my Sony digital camera. There is an orb present in one of the pictures, which seems to emanate its own light and contain a trail, which suggests that the orb is of a spiritual nature. However, keep in mind that the majority of orbs captured on digital cameras are the product of dust particles, lens refractions, moisture drops, or bugs, etc. I also acquired an EVP in this room of a female entity who spoke another language, possibly German.

I stayed in M001 for three nights during the *Queen Mary's* 80th anniversary of her maiden voyage. Cher Garman and I conducted some paranormal research in this cabin two nights in a row, especially after the ship quieted down after the festivities. At one point, Cher felt someone sitting down on the bed. More intriguing is that she actually saw the bed sheets move in a downward position! This occurred around 1:00 a.m. Earlier that night, I saw an ethereal figure appear right before me and move in a left to right fashion while I was reading a book in bed. I noticed that his clothing resembled a vintage black tuxedo.

As we conducted an EVP session in the servant's quarters of M001, we both heard the audible sounds of a young girl humming. We were able to document this on my TASCAM recorder as well. Additionally, we also heard a young female child who sounded very similar to Jackie. Since I have been doing research aboard the *Mary* for over ten years, I have learned to recognize Jackie's voice as she does have a very distinct cadence to her vocalization. So, when we asked if Jackie was present with us, we heard the girl say, "I love you." If you ever get the chance to spend the night aboard the *Queen Mary* and want to be in for a ghostly treat, I recommend staying in M001.

A024

I stayed in cabin room A024 on my very first overnight visit to the *Queen Mary.* A024 is located on the portside of A Deck, not too far from the hotel's front desk.

Several times, my friends and I would witness the sound of something banging the old-fashioned dresser furniture and knocking on the walls. I tried to come up with a natural explanation; however, I could not find anything substantial to suggest that these incidents were natural in origin.

Furthermore, I experienced numerous cold spots in this room. I turned the air conditioning off and the porthole window was shut. There was no natural explanation for these cold spots, which seemed to be ten degrees cooler than the regular temperature of the room. From what I have heard and read, A Deck staterooms seem to be very active with ghostly activity.

A192

A few paranormal colleagues and I made a recent overnight stay to the *Queen Mary*. We brought a lot of our investigative equipment with us to do a mini-investigation in the room. I recorded about four hours of audio in this room and acquired several EVPs. Many of the EVPs occurred while people were in the room carrying on with their normal conversations. Thus, it seemed as though the entities were intelligently interacting with us.

One of the investigators brought along his Ovilus and set it up in the stateroom. This device is used for real-time Instrumental Trans Communication (ITC) sessions. The dictionary mode was utilized in this gadget and the theory is that spiritual entities can choose which words they want to impart via the device. Another example of ITC work is evident when researchers use various spirit boxes, which are programmed to scan the AM and FM radio bands at an extremely high rate. Theoretically speaking, it is hypothesized that spiritual entities can manipulate the frequency of these boxes to formulate words or phrases in real-time.

Getting back to this particular evening, I left a recorder running in the room to record what the Ovilus was picking up. The Ovilus does not keep a dictionary of names and we ended up capturing many names that night. Some of them included *Larry*, *Caroline*, *William*, *John,* and *Bill.*

There were other key words that were captured on the Ovilus, which might directly coincide with the history of the *Mary*. The instrument kept spewing out words such as *run*, *fear*, *gun*, *remember*, *marine*, *brother*, *blood*, and *pain*. This could be indicative of the *Mary's* days during World War II. One of my sensitive friends also relayed that the spirit of a soldier frequented our room throughout the night.

A019

I was a featured speaker about the *Queen Mary* at Amy Bruni's ***Strange Escapes*** Conference aboard the ship on March 10th-13th, 2017. On Friday night (10th), I stayed on the starboard forward side of A Deck in A019. After the meet and greet, I decided to conduct a lengthy EVP session inside A019. In A019, I utilized two audio recorders—TASCAM DR40 and OLYMPUS WS-100—and synced the time on both, so their start times were exactly at 9:08 pm. Also, note that at 15:04 minutes into this session, I started to intuitively feel the possible presence of William Eric Stark in this stateroom.

We know that William Eric Stark was a former senior second officer who accidentally consumed tetrachloride instead of gin while aboard the ship. Sadly, he became very ill as a result and passed away a few days later. He is one of the documented crew member deaths who appears in spirit form aboard various areas of the *Queen Mary*. His spirit form has been seen and heard throughout the ship; he is often noted for his "grumpy" demeanor and deep/hoarse voice. I am familiar with the sound of his vocalizations, whether they are heard via EVP (electronic voice phenomena) or via the naked ear, as I have heard him on various occasions.

During this EVP session, I decided to do a little experiment with a former aft engine room audio capture of "Get out" that was obtained about two years ago during the late-night weekend Paranormal Investigation tours. During my A019 EVP session, I played this engine room "Get out" EVP and asked the energies if they know the person who said these words. I said, "Is this the voice of William Eric Stark or is it somebody else? If it's someone else, can you let me know the name of that individual?" There were no responses to these questions; however, it started to get really interesting when, at 24:07 recorder time, I captured an EVP in A019 from a male energy that 1) said "Get out" twice and 2) sounded strikingly similar to the aft engine room EVP played as a part of this experiment as well as the voice of Mr. Stark! Note again that these two captures are about two years apart. I do believe it's Mr. Stark's voice in both of these.

Also, at the time of my intuitive impression of Mr. Stark—at 14 minutes into the audio file—I said, "William Eric Stark, are you present right now? Are you in this room?" At :09—:10 seconds in the audio file, you hear a male whispered EVP response of "Yes." At 15:04 recorder time, I continue to say that "I am getting the impression of William Eric Stark." So, my intuitive impression of him inside A019 came about 10 minutes prior to the captured "Get out" EVP that was said twice. You can listen to these audio files on http://www.spiritedqueenmary.com under "Evidence Spotlight."

What's even more intriguing is that while on the following night's **Paranormal Shipwalk Tour** led by my friend and colleague, **Tony Ashlin**, about five guests heard Mr. Stark in R Deck Forward loudly saying, "Get out, out!" Since I am familiar with his vocal tone, I am pretty sure that it was him. At the time of this disembodied vocalization in R Deck Forward, I was not yet aware of the above mentioned A019 EVP as I had not yet reviewed my audio file from Friday night's EVP session. So, both nights (Friday and Saturday respectively), the highly possible voice of Mr. Stark saying "Get out" came through—one via an EVP in A019 and the other being a disembodied vocalization in R Deck Forward.

Was Mr. Stark somehow answering my question (was the "Get out" in the aft engine room two years prior the voice of Mr. Stark?) and letting me know it was him in the aft engine room saying "Get out" by saying the exact same words in the A019 EVP session? He could have directly answered my question and said, "Yes, that was me"; however, maybe he chose to say the words "Get out" as a unique way of letting me know it was him. If so, this capture shows the magnitude of spirit intelligence and communication. Then again, there is always the chance that this entire experience was coincidental (although I don't necessarily believe in coincidences) and/or the vocalizations of another male energy with a similar-sounding vocal tone.

AFT ENGINE ROOM

The *Queen Mary's* after engine room is famous for housing a hot-bed of paranormal activity. The spirits of J.P., a former captain, and a former engineer have all been documented. In my personal opinion, I believe that I have encountered the spirit of J.P. on several occasions. The most recent occurred in March of 2015 as I was in boiler room #3. As I was looking forward, I saw J.P. on my right side. He appeared with a smile on his face as he moved toward me, then backed up a little and dissipated. It was truly profound.

As I was taking one of the paranormal tours one day, I had an incredible personal experience while walking past water-tight door #13. I purposely wanted to be the caboose of the tour group. Thus, no one was traveling behind me. No one living that is! After the tour guide was finished educating the tour group about J.P.'s accident and his ghostly appearances, the group proceeded to walk through the door to another location in the engine room. Again, there was no one behind me. Right after I walked through door #13, I felt the distinct tug on my purse from unseen hands. I ruled out the possibility that my purse brushed up against the railing because I was walking in the center of the aisle. I remember turning around after I felt the tug on my purse to find that no one was present.

I have captured several EVPs in the engine room. One of my engine room audio anomalies stands out among the rest and I feel that this particular capture is the spiritual voice of J.P. This EVP actually occurred in close proximity to door #13. On another tour, the guide asked the group if anyone would like to share any paranormal experiences that they have had in the engine room. Of course, I eagerly raised my hand to tell of the time that I had my purse tugged from invisible hands. I had my recorder running during this time. This is what occurred: I proceeded to tell the tour group what happened. I said, "On another tour in the engine room, I was walking through door #13 and felt something tug on my purse, but there was no one there." After review and playback of my audio, you hear a young male's voice say "I am down here" right after I finish saying "...but there was no one there."

In my opinion, this entity was intelligently interacting with me to let me know that I was in fact wrong and that there was someone with me. I believe that this EVP was J.P. communicating with me and letting me know that he was there. Whenever I am down in the engine room, I always feel as if unseen eyes are watching me. This feeling always intensifies around water-tight door #13. Other people have reported this feeling as well. I elaborate more on J.P. in *Spirited Queen Mary: Her Haunted Legend.*

FORMER FIRST- AND THIRD-CLASS POOL ROOM

The former first- and third-class pool room is by far the most talked about area of the ship. Many people believe that this area is the most haunted location on the ship. Scores of people come through the pool room each day on various tours, thus indicating why it is purportedly said to be the most paranormally active area of the *Mary*. Furthermore, psychics have located what they believe is a portal or vortex in the pool area's dressing rooms. A portal can be described as a passageway for spirits to travel to and from this world.

I have had numerous paranormal encounters and experiences in the pool room, which have been felt by all the human senses. I will share with you some of my unique experiences while visiting this area of the *Mary*. As some of you might already know, the *Mary's* most famous spirit, Jackie, is the most popular ethereal being experienced in this spot. Many people of all ages have reported feeling, seeing, or hearing her. I am one of those people. I have had the pleasure of capturing some amazing EVPs of a female child in the pool room. I have also seen the floating face of a small female child in this area as well. Jackie is said to resemble the legendary child actress Shirley Temple.

I had my recorder running during one of the tours, mainly to record the content of the tour. However, I also wanted to see if I could maybe catch some interesting EVPs as well. Since there were about ten people in the group, I was very meticulous at documenting erroneous sounds into my recorder. There were no children present or adults with child-like voices in the pool room on the tour.

During this particular tour, the *Queen Mary* still had its ghost camera up and running where people from home can watch the camera in hopes of glimpsing something paranormal. My mom was at home watching the camera via the internet as the group and I entered the pool area. At one point, I remember saying to one of my friends, "My mom's watching us on the ghost camera." Upon review of my audio, you clearly hear the angelic voice of a small female child say "My mommy" right after I finished telling my friend that my mom was watching on the device. What is interesting about this experience is the fact that I mentioned the word "mom" and the EVP contains the word "mommy." I do not find that to be coincidental. About 20 seconds or so after this EVP, you clearly hear another EVP of what sounds like a small girl giggling. Both of these experiences have been reported by other people as well.

On another trip to the *Queen Mary*, I had the luxury of visiting the pool room late at night with some paranormal colleagues. It was during this time that I saw white mist float near the ceiling and the ghostly face of a small female child, which seemed to be floating toward us. She looked to be about five or six years-of-age. She had fair skin, brown eyes, and curly hair, which came down to her shoulders. She did not seem to have any expression on her face and my glimpse of her lasted a fraction of a second. It was an experience that I will never forget!

During this time, my colleagues and I also experienced the sound of banging, which seemed to emanate from within the walls of the pool area itself. We were doing an EVP session and some of the banging seemed to coincide with our questions. At the time, we could not rule out whether the sounds were due to the ship itself. One of the investigators reviewed her audio piece and found that she had an EVP of a small female child as well.

On another trip to the *Mary*, some investigators and I were in the pool area when I acquired an interesting piece of audio. I was telepathically asking Jackie to make her presence known to me. A short time later, I acquired an EVP of a small girl saying what strongly sounded like, "Jackie." This voice matched all of the previous EVPs of Jackie that I have obtained. Since I am an educator, I have a background in phonetics, and phonetically, you hear what sounds like the aforementioned name. If this was in fact Jackie saying her name, was she responding to my telepathic communication?

Another piece of amazing audio that I have in my pool room EVP arsenal contains a series of different EVPs within a matter of seconds. This audio segment contains a disembodied voice yelling "Hello," followed by the sounds of wet footprints and splashing water. The segment ends with a ghostly voice saying "Slide in" in reference to the pool's slide. Remember that the pool has been drained of water for many years.

On a separate visit to this hot spot location with another tour group, I captured what sounded like an elderly female. What is unique about this piece of audio is that I felt like I was being watched and uninvited at the time the ghostly voice occurred on my recorder. When I reviewed the tape, I heard an EVP of an elderly woman saying "Hello" in an angry tone.

I was at the *Queen Mary* on the night of December 29, 2009, with some friends and fellow paranormal colleagues. When we were in the pool room, I stationed myself on the second portside level when I noticed a shadow figure move at the opposite end. I audibly let everyone know what I saw. As a side note, I was recording our time in the pool room. A deep male voice said "Hello" about two seconds prior to me informing my friends that I saw a shadow. Furthermore, I proceeded to ask more questions. I asked who was in the pool with us and what his name was. As I played back my audio, I heard the distinct male response of "Andrew" right after I asked the question, "Can you tell us your name?" The response was not audibly heard at the time of recording; thus, making it an electronic voice phenomenon.

FORWARD CARGO HOLD PIT

I have had the opportunity to visit the *Queen Mary's* Forward Cargo Hold on two separate occasions. The Forward Cargo Hold (FCH) was used during World War II to house Italian and German prisoners-of-war (POW). It has been said that the POWs had standing room only because they were sadly packed in the area like sardines. They were only given bread and water, which was lowered down into the area by a bucket. Many of these prisoners took their last breath in the FCH and their bodies were unceremoniously thrown overboard into the sea. Thus, it is no wonder that this area is one of the most paranormally active, if not *the* most active area of the ship.

In order to actually get into the FCH pit, one has to climb down this narrow metal stairwell. It's very dangerous, and as of now the ship does not permit tours to enter the location. To this day, I have investigated this spot three times. On the very first visit to this daunting area, I was immediately overcome by the leftover emotions of the prisoners-of-war. I felt that I was constantly being watched by unseen eyes.

During a late-night paranormal tour in the FCH pit, the group decided to do a controlled EVP session. During this session, I heard an audible groan, which seemed to emanate from someone who was very weak and exhausted. It was clear that this vocalization did not come from one of the tour members. Upon playback and review of my audio during this session, you clearly hear a ghostly male voice say "Hungry" right after I ask the question "Is there anything that we can do for you?" I must say that goose bumps popped out all over my body when I heard this EVP. Tears also enveloped my eyes as my heart was overcome with emotion. To date, this audio capture is one of the most heartbreaking spirit voices that I have ever heard.

On another trip to the Forward Cargo Hold, there were many distinct taps against the ship's metal, which seemed to coincide with the questions that were being asked. One of the taps even sounded like Morse code. Furthermore, I heard the sounds of rustling footsteps occur in the area right above the FCH pit. Many people don't realize the amount of ghostly activity in the FCH simply because it is a less common area to visit on the ship. If you ever get the chance to visit the *Queen Mary*, I really hope you have the opportunity, assuming it opens up again, to travel down to the Forward Cargo Hold.

BOILER ROOMS

The former boiler rooms are also noted as having paranormal activity from time to time. One night, some investigators and I visited the Green Room. When we were inside, a young male entity was communicating to us and his presence was sensed due to the drastic cold spots that were occurring. A gifted sensitive affirmed the fact that a male presence was in the room with us. Perhaps this was John Henry greeting us.

Right outside the Green Room, an investigator was trying to get in touch with the spirit of a former *Queen Mary* captain. On many occasions, a distinct tapping against metal could be heard right after the investigator's questions. This may or may not be coincidental, but the tapping was rhythmic and suggestive of communication. *Spirited Queen Mary: Her Haunted Legend* delves more in detail as to the ghostly activity inside these areas.

PROMENADE DECK

In January of 2010, my mom and I were browsing the various gift stores prior to walking on Sun Deck. Little did we know that we were about to have an eerie experience while we ascended onto Sun Deck via one of the stairways. As we were on this particular stairway, we both smelled a very distinct and strong sense of cigar smoke. There was absolutely no reason for the smell's existence. There were hardly any people around and out of those that were present, no one was smoking—or smelled of smoke for that matter. The phantom cigar smell was isolated on the stairs. We tried to come up with a rational explanation for what our noses were picking up, but there was no such logical reason for it. Was the smell emanating from the ghosts of Winston Churchill or William Eric Stark as they took their daily stroll? Or perhaps it was a residual aroma left over from the World War II days when soldiers occasionally smoked.

In the following photograph captured by Amy Liam McCallum, you will see an apparition of a headless man adorned in a suit walking forward on the port side of Promenade Deck.

SUN DECK

The *Queen Mary's* Sun Deck is very well traveled simply due to the scenic ambience it offers. I had a very interesting experience one night while walking on the Sun Deck on the ship's portside. I felt water splashing onto my face. I must note that the weather had very little humidity that night and there were no clouds in the air. No one was around me that could have spilled water on me. Thus, I could not figure out why this occurred.

I later researched this to see if other people have had this occurrence. I was surprised to find out that this is a common anomaly that is reported. The *Queen Mary* had to travel some very rough ocean during its sailing days and has experienced sailing through severe storms. Thus, it is no wonder why some people have experienced this residual phenomenon.

THE CAPSTAN'S CLUB

The Capstan's Club is located at the ship's stern on A Deck. It was used for parties, banquets, and meetings. Now it serves as the Wyndham Preview Center. It is another area aboard the ship where I have experienced and acquired some incredible ghostly evidence. I was staying on the *Queen Mary* overnight with a few paranormal colleagues and friends. Three of us decided to go into the Capstan's room around 1:00 a.m. All three of us are sensitives and all three of us experienced activity in the room.

During an EVP session, we all heard the distinct sound of someone shuffling their feet. However, no one was present inside the room besides the three of us and we were stationary at the time. It was early in the morning and no one was outside walking the decks either. We could not come up with a logical explanation for the footsteps.

A little later during the EVP session, one investigator started to empathically pick up the emotions of a young soldier. He kept saying that he was sensing a young soldier who was in despair and missing his mother. I also had my recorder running during this time and I picked up a few anomalous male voices. This particular researcher was the only living male in the room at the time. What he was feeling was so strong that it brought him to tears and this empathic moment lasted about five minutes.

I was at the *Queen Mary* on the night of December 29, 2009, with some friends and paranormal colleagues. After we visited the pool room, we decided to go to the Capstan's Club. In this area, we conducted a short ITC (real-time communication) experiment before doing a short

EVP session. One of the questions that I asked was, "Did you fight in World War II?" Upon playback of our audio, some of our recorders, including mine, had a male response to my question. I couldn't quite make out the response, but another friend deciphered it as answering, "Yes I did" in reference to my question.

SUBMITTED STORIES: ENCOUNTERS WITH THE PARANORMAL

PATRICIA V. DAVIS

Best-selling Author, Cooking for Ghosts*, www.TheSecretSpice.com*

My paranormal experience aboard the RMS Queen Mary was unique, for the reason that—I'm embarrassed to admit—I knew little about the ship when I first stayed aboard—little about her history, and nothing about the rumors that she's considered haunted. I was in Long Beach on an assignment at an October women's conference, and the convention center where the conference was being held was already booked. So, I went to hotels.com and found the best next option: The Queen Mary.

I'm astounded now by how much I was oblivious to—the beauty of the ship, the number of people walking around with guidebooks—all of it, because my focus was on the job ahead of me the next day. That was why when my glasses moved around in my stateroom, not once, but three times, to three different locations, the last from my dresser to my bed pillow, I really wasn't thinking "ghosts," I was thinking that I was stressed over the conference and being absent-minded.

But there were other things: footsteps echoing as though someone was walking in the room with me, a sense of melancholy or foreboding in certain parts of the ship while I was going down to dinner, but most of all, there was that movement of my glasses. It seemed almost playful. After the conference, when I was flying back home, I picked up the in-flight magazine in the seat pocket in front of me and started flipping through it. As this was October, this edition was all about harvest celebrations and Halloween.

One article that hit my eye: "Top Ten Most Haunted Places in the United States." And who was on the list? The Queen Mary, *of course. When I read about a mischievous little ghost thought to be the spirit of a little girl who'd drowned aboard the ship many years before, my fascination with the ship was born. I read everything I could find about the majestic ship and her ghostly residents, including a little girl. A little girl who enjoys filching people's eyeglasses.*

ROSEMARY ELLEN GUILEY

Paranormal Researcher and Author, Visionary Living, Inc.
Rosemary Ellen Guiley is a well-respected paranormal researcher, lecturer, and author in the United States and throughout the world. She has written numerous books on the subject and has appeared in numerous media outlets regarding the paranormal. I had the pleasure of meeting Guiley in June of 2009. Both Rosemary and I are members of the Ghost Research Society and have researched together at Waverly Hills Sanatorium and also aboard the *Queen Mary*. In her own words, Guiley shares some of her paranormal experiences while visiting the RMS *Queen Mary* during a TAPS sponsored event:

My most remarkable experience of the haunting phenomena aboard the Queen Mary *took place at the famous door #13 below decks, where a crew member was crushed to death. He was an 18-year-old crewman who was killed when he was caught in water-tight door #13 in 1966. During a TAPS event investigation, I was with a small group of people near door #13. I was standing a bit off to myself when I felt a sharp tug on the back of my jacket. There was no one behind me. A few minutes later, I caught a fleeting glimpse of a young man in overalls, which vanished near the door. At the time, I did not know that an apparition of one of the killed crewmen was seen there. I wondered if he was the one who pulled on my jacket—a common ghostly phenomenon in haunted places. Why do they always approach you from behind? Is it to play the trickster, give you a jolt when you turn around and see no one there?*

In the infirmary, I heard the murmuring of many voices. The atmosphere there felt heavy, oppressive, and tense—and also somewhat sad. I experienced a similar scenario in one of the holds. Peter James once said that he contacted more than 150 ghosts on the Queen Mary. *There does seem to be quite a few residual haunting effects, some vague and some strong, almost anywhere you go on the ship. I have been to the* Queen Mary *three times and consider it quite haunted. While I consider the* Queen Mary *to be quite haunted, its phenomena were not as strong for me as some other famous haunted sites I have investigated.*

PATRICK WHEELOCK

Investigator, Beyond Investigation Magazine, *KBIM Paranormal Talk Radio*
Patrick Wheelock has been investigating the *Queen Mary* for paranormal occurrences for many years. His method of investigating is strictly scientific and *Beyond Investigation Magazine* approaches its paranormal research and investigation very meticulously. He previously conducted paranormal tours on Friday nights aboard the ship. I have had the honor to be a guest on his tour a couple of times. The following are three of his profound encounters aboard the ship:

There was a relatively small group that night, roughly fourteen people or so for the tour. As usual, we had started out in the Paranormal Research Center for the briefing and headed down to the first class pool area. The group spent fifteen or twenty minutes in the first class pool area and then we headed forward and down into the Forward Cargo Hold. The initial readings in the pit were nominal and the group got settled in for a round of voice recordings or EVPs. One young girl (approximately fourteen years-of-age) sat to my right with no one on her right. The rest of the group sat across from us. The group decided they wanted the lights turned off for the recordings and all the flashlights were turned off.

We began the recording process as normal with a series of questions with delays in between to allow any unheard voices to answer. A short time after we started, the young girl to my right called out that someone was touching her back. We turned on the flashlights and made sure that no one had moved near her to be able to touch her. After confirming she was not scared by what she experienced, the lights were once again turned off. A minute or so after the lights were turned off, the same girl called out that someone was pulling her hair. At the same moment, one of the guests across from me called out that the temperature had dropped from 68 to 57 degrees. We turned on the lights and sure enough, the girl's ponytail was sticking straight up as if someone was holding it up. The hair stayed in that position for just under two minutes and then fell.

When her hair fell, the temperature was reported to have jumped back up to 67 degrees. I checked to make sure the girl didn't have any thread or wire causing her hair to stand up that way, and static readings the entire time were relatively low, indicating static electricity was not the force causing the hair to stand up. In my five years aboard the ship, that is the first time I have ever seen something that significant with my own eyes.

Patrick Wheelock shares another experience, which occurred in the former first- and third-class pool room. *Beyond Investigation Magazine* thoroughly examined this occurrence and has not been able to pinpoint a natural explanation for it. Here is the story in Patrick's own words:

The Beyond Investigation Magazine *(BIM) Team was conducting a 72-hour investigation of the pool area in 2006. We had multiple cameras and microphones feeding back to the base console at the end of the long hall leading into the pool area. On the second night, a little past 3:00 a.m., we recorded the sound of a child's giggle. No one on the team heard it when it was recorded and we didn't discover the sound until months later.*

We began a series of tests on the recording as well as filtering to try and make it clearer. The high pitch of the voice didn't match any of the BIM team members' voices, so we knew we were looking for another source. We took the recorded giggle and copied it to a CD player. We placed the CD player at different locations outside the pool area. Two years later, we found one possible answer for how the giggle was recorded on our equipment. We found that if we played the recording in a very particular location near the elevators by the main entrance to the pool, turned the player to a very specific direction and to a specific volume setting, the voice traveled just enough for our recorders to pick it up and not be heard by the team.

Now, the question is, do we really think that, at a little past 3:00 a.m., there was a little girl near the elevators giggling? More than likely not, but as an investigative team, we must search out all of the possible explanations. Furthermore, the story of a little girl drowning in either the first- or second-class pool has no merit. We have been unable to find any verifiable forensic evidence that anyone, child or adult, drowned in either pool, which I guess makes the giggle that much more of a mystery. If there wasn't a real child at the elevators giggling, and there was never a drowning in either pool, then what caused the recording? What is causing an almost daily flow of reported incidences with a small girl in the pool area?

PATRICK WHEELOCK'S THIRD ACCOUNT

For about two months during one summer, the tour groups, my cameraman Anthony, and I witnessed something, that to this day, I have been unable to determine the cause of with any logical validity. As we investigated the cargo hold itself, we had split the group in two to allow everyone to enjoy the tour.

We were on the port side of the cargo hold in the ceiling area when I caught something in the corner of my eye. I quickly turned my head just as a woman in the group said 'Oh my God, what's that?' The rest of the group turned and saw what I can only describe as a lightning bolt or plasma arch, which was bright, but did not light up the area. It was moving as the typical plasma displays do, with no pattern, simply squiggled and ever-changing. The light moved from the center wall, slowly toward the port side, until it reached approximately halfway, where one of the original light fixtures was mounted, which took on average of 4.2 seconds. Originally, the light would repeat after a few minutes, but after witnessing it twice, it would only happen once each night during the tour.

There are several small holes in the aft wall of that area and to look through them, one can clearly see pipes, wires, and the ship's original water softening plant. If you follow the original light fixtures, the wiring conduit runs into the water softening room. After the second time seeing the light, I had security let me into the cargo hold area one night with a ladder and an equipment case. I tested the original light fixtures in the ceiling area, both port and starboard side. Those fixtures did not have power to them at all, not even low voltage. The two bow end light fixtures did have power, but both bulbs were burned out. I then connected a 100 ft. extension lead from one of the bow lights to make sure that the bow fixtures still had power, when I re-tested the aft light fixtures. My second test proved that the bow fixtures only had power to them. Being that the aft fixture wires ended into the water softening room, I had to make my way into there to see if and what they were connected to.

After spending a few hours trying to squeeze my big belly into tight quarters within the water softening plant room, I came to the conclusion that aside from me taking apart the softening plant itself, I wasn't going to be able to trace down those two sets of wires that feed the two aft light fixtures. Unfortunately, that put an end to, once and for all, determining if those two fixtures ever receive power. The fact that the ship is currently wired in three ways for power, each of which is not compatible with the other, adds to the plausibility of some sort of power issue causing the light.

The light was witnessed, as I stated earlier, almost every week for at least two months. I would estimate that 150 or more tour guests witnessed the light. Strangely, when pictures were taken with digital cameras, the images always came out as a white spot with no defining shape and definitely not like the lightning or plasma bolt it looked like as we watched it happen. All video footage came out as completely white, as if we were shooting a bright white piece of paper. Now, being completely color blind, I cannot tell you that I saw the light as anything other than white, but the guests and Anthony stated the light was "blue" in color. Sound was never disrupted and none of the investigative equipment picked up anything strange during any of those times.

One other note: I made a point of being in the hold during the same time period on different nights of the week to see if the light effect would happen. On the numerous times I went in on off days, the light effect never happened, which would also give more credit to the possibility of a power issue, that is on a timer or bad relay causing the phenomenon. I had planned on and received approval to spend two nights in the hold with multiple cameras broadcasting every corner of the hold, live, via the BIM website, but a management change aboard the Queen Mary *ended the project. After the roughly two-month period, the light effect stopped and as far as I know, has not been seen again.*

To this day, I still do not have a verifiable explanation of what caused the light. I guess if I ever do make it back to investigate, I would have to return to the forward cargo hold to try to find an answer to the light phenomenon.

ERIN POTTER

Paranormal Investigator, Paranormal Housewives

Erin Potter is an investigator with The Paranormal Housewives. She and I have conducted paranormal investigations together aboard the *Queen Mary*. Here, she shares a few of her paranormal experiences on the *Mary*:

We were in the first class pool and we were doing an EVP session. Questions were being asked and Nicole said, "There is a little girl here named Jackie." I then got an EVP which was a distinct little girl's voice saying "Hello." The voice sounds the same as all of the other pieces of audio containing what Jackie's voice is said to sound like, whether they are audibly heard or caught on voice recorders.

That same night, Leif and I got back to our room about 2:00 in the morning. I had turned on my recorder and set it on the table by the bed. Leif and I went into the bathroom to get ready for bed. We then went to bed and the recorder was on the whole night. When I played back the recording the next day, what I heard was completely awesome. In the first ten minutes of the recording you hear the television on in the background.

You could also hear Leif and I getting ready for bed in the bathroom talking about the night. About five minutes into the recording, you hear someone snoring really loudly. It was so close to the microphone that you could actually feel the breath on the microphone through the recording. It happened again a few minutes later and then it didn't happen for the rest of the recording. Leif and I were alone in the room. You can hear us in the bathroom so we were not near the recorder. You also hear the toilet flush, which is not the same sound. It was an actual snore by what seemed like an old man. You can even hear his lips flapping as he exhaled.

A COLLECTION OF PARANORMAL STORIES FROM SHIP VISITORS

Courtesy of *Queen Mary Archives*, the following documented stories are paraphrased in my own wording. For confidentiality purposes, first names are only used. For creative purposes, I added a title to each story.

THE SPECTER OF A170

On August 8, 1991, some friends decided to visit the *Queen Mary*. They stayed in cabin room A170. One of the guests described how she had a very hard time going to sleep because she felt as though someone was watching her from the other side of the room. She relays that when she sat up in bed, she saw a blue light coming from the corner of the room. In that blue light was the ghostly figure of a little boy in blue who looked to be in his early teens. The ghost looked as though he was smiling and the guest said that she had a warm feeling come over her at the sight of the ghost. The next day at breakfast, one of the guest's friends relayed that he had seen the specter of the boy as well.

A STATEROOM ENCOUNTER

A guest describes her encounter that took place in October of 2003. When the guest turned off the light in her stateroom, she felt as though another presence was in the room besides her and her husband. As she lay down in bed, she could feel that her blankets were moving and that something touched her foot. Needless to say, she did not sleep that well that night. The next morning after the guest took a shower, she noticed small scratch marks on her neck. She did not know how they got there.

THE WILLOWY SPECTER

A mother and daughter were visiting the *Queen Mary* with some family and encountered the ghost of an older woman. The guest relays that while they were on the ship, her mother saw a lady dressed in a long dark coat with a fur collar, dark stockings, and dark dress shoes. The female specter had shoulder length hair and appeared to be around 5 foot 6 inches tall. This encounter took place in an area that displays the soldiers' quarters. The ghost was described as entering the room through the left wall.

DRESSING ROOM GHOST

A former Ghost and Legends employee describes her encounter that took place in the pool area's dressing room. This former employee arrived at work one morning quite early. As she was in the dressing room, she felt like someone was behind her, and as she turned around to see who it was, she saw the ghost of a young girl in a bathing suit. The young ethereal energy said, "You're not my father," and walked away. As she walked away, the apparition faded into nothing.

MULTIPLE APPARITIONS

A family reported numerous ghost sightings while visiting the *Queen Mary* about five years ago. The family went on one of the tours and encountered disembodied voices. As they entered one of the former boiler rooms, one of the family members had an empathic experience where she felt the pain and suffering associated with those who were burned to death. Furthermore, during this empathic experience, the guest reported that she saw two men and a woman being burned. The sighting of a man in a dark blue uniform with gold buttons was also reported. Apparently, the ghost in the uniform was following one of the family members for awhile.

SHADOWS IN THE BATHROOM

A couple decided to stay on the *Queen Mary* in October of 2003. The couple had a paranormal encounter while staying in one of the staterooms. When they were getting ready to go to sleep, the wife reported that she saw shadows in the bathroom moving around. Later on, while she was in bed, she continuously felt the bed sheets moving and the feeling of someone touching her foot. This person also reported scratch marks on her body as well.

CABIN 428 EXPERIENCES

A guest by the name of Sally describes her experiences on the *Queen Mary* back in 2002. She stayed in an authentic stateroom, meaning that the cabin has been preserved to look like it did during the ship's sailing days. She stayed aboard the ship for a few days.

In Sally's authentic stateroom, she noticed the door knob moving back and forth of its own volition. She left a cassette recorder in her room to see if she could capture any anomalous voices on tape. Upon playback and review of her audio, she heard a screaming male voice and the voice of a female who was singing in a foreign language. Whispering male voices were also heard on her tape.

Sally took one of the late Peter James' paranormal tours. While on the tour, she was tapped on her shoulder by unseen hands. James also relayed to the tour group that many people have reported their phones to ring only to find that no one is on the other end. When Sally had to get up early to catch a flight, her phone rang at 6:00 a.m. No one was on the other end. She later asked the front desk if she had received any calls. To her surprise, the front desk agent's response was "no." This has happened to me on various occasions while I stayed aboard the ship for a couple of nights.

Sally also relays what she saw on one of the *Queen Mary's* ghost cameras. The ship had ghost cameras set up in various places so that people from home could monitor the cameras to see if they could find anything unusual. Well, Sally reported that she saw what looked like a male ghost in one of the dance halls. She described the specter as being enveloped in a light blue light. She did not notice his facial features, but did notice that he was wearing some sort of work clothes. Perhaps a boiler or engine room crew member?

THE POOL ROOM APPARITION

A few family members visited the *Queen Mary* about three-and-a-half years ago. As they were taking a tour of the ship, they made sure to have their video recorder running. As they entered the former first- and third-class pool room, they caught something very strange on their camera. The strange man with dark hair was wearing what looked like black and white striped prison clothing. Apparently, this ghost walked along the left-side pool deck under the balcony and disappeared. Thus, he never reappeared on the other side of the column that was supporting the balcony. The image of this ghost is clear and on tape for about two seconds. None of the other tour members saw or mentioned anything about this particular specter.

THE WINDSOR SUITE SPECTERS

A guest describes her experiences in the Windsor Suite on November 8th and 9th of 2003. The guest mentions that she kept the light on in the servants' room due to the fact that she was afraid of sleeping in the dark. She was waiting for her other friends to arrive on the ship, so she decided to go to sleep early. As she lay in the bed, she heard the front stateroom door close by itself. She figured that her friends had arrived. About 15 minutes later, she felt as though someone was tickling her feet. The guest mentioned that she made a loud moan to make the person stop tickling her feet. She then heard a voice say, "Leave her alone; let her sleep."

The following morning, room service woke the guests up. As they ate breakfast, they talked about the night prior. The particular guest mentioned that she did not appreciate her friends coming in her room and tickling her feet. The friends looked shocked as they were not in her room the night before. Needless to say, the visitor moved out of the servants' room and slept in another portion of the suite.

THE "OLD SPICE" GHOST

A former *Queen Mary* employee shares her experience with the famed "Old Spice" ghost that she had back in 1999. During her first week of employment, as she was walking down A Deck towards the employee elevators, she suddenly became engulfed in the smell of Old Spice cologne. She started to look around for any broken cologne bottles, thinking that someone dropped a bottle on the floor. She did not discover any such bottle. In this particular area, there were no bedroom suites, doors, openings, or people nearby that could account for the pungent smell of Old Spice.

Less than 20 minutes later, the former employee returned to the same spot, only to find that the smell had completely disappeared. Her co-workers then told her the tales of the ship's "Old Spice" ghost. This ghost has been sensed in various areas of the ship. I have had the experience of encountering this particular specter on many occasions. Specifically, I have experienced him while on A Deck.

M210 TALES

On New Year's Day, a group of friends was unpacking their bags in M210. As they started getting their cameras ready to take pictures, one member motioned for the other friends to come to where he was standing. Apparently, there was a bundle of red hair sitting right next to a battery. None of the guests had red hair and they could not figure out where it had come from. They tried to come up with a logical explanation and could not arrive at any such explanation for the red hair!

Around 5:30 a.m., the guests reported that they heard a loud bang in M210. They described it as sounding like large wooden tables being dragged across the floor. The sound also mimicked the ship scraping against rock. The sound seemed to circle around the entire stateroom. This noise could have originated from the ship; however, the guests again tried to come up with a logical explanation and could not arrive at any logical conclusions for the weird sounds.

GHOSTLY ATTRACTION

A guest describes her ghostly encounter that took place on the ship back in the 1980s. The guest and her husband went to the *Queen Mary* to celebrate New Year's Eve. The wife went out to the deck to enjoy the scenic surroundings when she heard a ghostly voice say, "You're beautiful; would you like to dance?" The woman described the apparition as having dark hair and wearing a tux and top hat. She apparently jumped at the sight of the ghost, and it just disappeared right before her eyes.

A GHOSTLY "HELLO"

A woman and her 13-year-old son had an interesting ghostly encounter while visiting the ship in January of 2003. The two were pretty hopeful to experience a ghost, especially since they came fully equipped with equipment known for capturing paranormal activity. They have visited the ship on numerous occasions.

In the hallway on A Deck, the boy captured Electronic Voice Phenomena (or EVP). Upon playback and review of the audio, they captured the audible and cryptic voice of a woman saying, "Hello, Hello, Keep out, Keep out!" A few days later, the woman and son went back to the ship to do more investigating, and they went back to the same spot where they captured the above voice. As their recorder was running this time, the ghostly woman's name was asked. A few seconds later, the boy's mom felt a breath and heard the disembodied voice of a woman whisper in her ear. She asked her son if he had heard the whisper and he said he did not. Additionally, both mother and son smelled a very fragrant perfume in the same location in the hallway. No one was present but the two of them. The phantom smells of perfume and cologne have been sensed by many people onboard the ship.

A DISEMBODIED VOICE

A guest describes the unique experience that he had while visiting the ship. Around midnight, the guest and his friend were getting ready to go to bed when one of them noticed the chain on the stateroom door had been fastened. When one realized that the other had not put the chain on the door, they both pondered who could have done it. A recorder was also running while they were sleeping.

The next day as they listened to the previous night's recordings, a man's voice was recorded, saying something to the effect of, "I feel home or I feel old or I feel alone." There was no man in the room with the two friends the previous night. The voice was described as definitely emanating from inside the stateroom itself.

THE DISFIGURED ENTITY

An adult describes his experience on the *Queen Mary* back when he was a very young child. He relays that he remembers it clearly. When he was a young boy, he remembers walking down a corridor on the ship that leads into another room. He began to feel a presence and turned to his left, only to see the mangled body of a man. The ghost was described as missing his left arm from the elbow on down. There were parts of the disfigured ghost's skin missing. To see this at a very tender age is indeed frightening. He remembers screaming at the sight of this apparition.

THE GHOST OF A BEARDED MAN

Two family members stayed aboard the *Queen Mary* in the summer of 2000. They describe how they were joking about the ghosts of the ship while taking a stroll through the engine room. They kept telling each other how fun it would be to encounter one of the ship's spirits. Their wish eventually came true.

As they were inside the propeller room, an eerie silence startled them. One of the family members described the cold spots engulfing the room and the sound of disembodied footsteps. The footsteps grew louder as the ghost of a bearded man in baggy clothing was seen. The family members said that the apparition was walking towards them. They frantically went up the stairs and looked back and the ghost had disappeared. Other tourists have seen the specter of a bearded man as well.

ENGINE ROOM ENCOUNTERS

An employee describes her experience in the engine room when she worked on the Shipwreck attraction. When this employee had her paranormal encounter, she was all alone in the engine room. She apparently saw two white shoes walking her way. At first, she mentioned that she thought it was her friend walking down the corridor. She also saw shadows drifting around the area as well.

The same employee recalls another paranormal experience, which occurred the following day on the ship. The employee was doing her job in the engine room when she said that something struck her shoulder quite hard. There was no one around who could have done this.

FIRST CLASS ENCOUNTERS

A tourist recalls her experiences in one of the first class staterooms. The closet doors opened and closed of their own volition. The bed sheets were reported to lift by themselves. The stateroom door apparently locked by itself. Then the tourist describes seeing the ghostly sight of a skinny blond-haired woman who was wearing 1930s clothing. The apparition was said to have disappeared through a wall.

BRITANNIA SALON SPECTERS

Some family members were staying in cabin room M160, which is right near the *Britannia* Salon. As the guests approached the doors to the *Britannia* Salon, a flash of white light was seen moving across the floor. This family also mentioned that they captured some ghostly photos as well during their stay aboard the *Mary*.

THE BAR GHOST

In 1999, a wife and husband were at the *Queen Mary* for a New Year's Eve party. The husband relays that after dancing with his wife, he wanted to go outside to get some fresh air. As he was outside on the deck about to light his Cuban cigar, someone tapped him on the shoulder and said, "Can I bother you for a light?" The husband said "yes," and when he turned around to light the cigar, no one was there. He scanned the entire deck for the person who asked him for the light, and that person or ghost could not be located.

LADY IN GREEN

A woman by the name of Barbara describes the paranormal experiences that she had when she went to the *Queen Mary* back in 1988 with a friend. Specifically, she shares the encounter that she had in the cocktail lounge. When her friend went to the restroom, Barbara looked over to her right and immediately saw a woman sitting three tables away. Barbara describes the female ghost to be quite beautiful, wearing a solid green antique satin Victorian dress. She also wore a large matching satin feathered hat and parasol. The ghost had dark hair that was pinned up and she appeared to be in her twenties. The specter sat at the table for three minutes as Barbara just stared at her. At the end of the three minutes, she glanced to see if her friend was coming out of the bathroom, and when she looked back, the young ghost was gone. Barbara knew that the apparition did not have ample time to have walked out of the room in the time it took her to glance away for a second.

Barbara relayed that after this experience with a *Queen Mary* ghost, her concept of the world changed. She said, "I believe your ship is a special place where the souls who long to relive their happiest moments, still have that place to go to. I am thrilled and humbled that they chose to allow me to come into their world."

SUITE ENCOUNTERS

A newlywed couple stayed in one of the suites aboard the *Mary* in 2001. The couple actually got married on the ship and spent their honeymoon or a portion thereof on the liner. They did not realize that they would be in for a ghostly treat.

At one point, the elevator decided to open up on its own accord. This is another common occurrence that is reported frequently. About midnight, the couple went to their suite. During this time, another relative who was there celebrating found himself locked in the bathroom, even though he did not lock the door himself. Who locked the door? The female guest reported that a party was going on above their stateroom at 3:00 a.m. She reported hearing sounds of music and high heels tapping on the floor above. The next day, she inquired about the dancing. The hotel manager said that the rooms above the couple were empty due to renovations.

This female guest has relayed what I often feel when I first walk on the ship. This is what she had to say: "Ever since I stepped foot on the ship, I have felt an immediate connection to her, as if I was on the ship before." This feeling has been felt by crew, passengers, and visitors from the dawning of the *Mary's* days.

M001 SHADOWS

In April of 2003, a paranormal occurrence took place in M001. A guest reported that a door shut and that he had seen shadows of someone walking right past the two beds and into the corner. After looking around, he then found the closet door ajar. Since I have stayed in M001 myself, I can attest to the fact that ghostly events do take place in this cabin. I wonder if the M001 apparition I saw in May of 2016 was the same specter that this guest encountered when he stayed in this cabin?

GHOST CHILD

A husband and wife were taking a tour one day when they came into contact with one of the ship's ghost children. They were in the pool area when they heard the sound of water splashing and the sound of a child having fun. They saw wet footprints as if someone just got out of the pool. The pool's water has been drained for quite some time.

ENGINE ROOM WHISTLES

A passenger describes her experience, which took place in the engine room a few years ago. She said that as she was standing in one of the engine room's walkways, she heard the distinct sound of someone whistling. She was all alone at this time. She looked over the railing, and to her surprise, she saw a young man with a full beard who was wearing overalls. The ghost had his hands in his pockets as he was walking around. The guest said that she turned to walk away and noticed that he was about two feet from a wall. When she turned back around, the ghost had simply vanished.

GIFT SHOP POLTERGEIST

A sheriff's detective describes his experience that he had, even though he did not believe in ghosts at the time. The detective and his daughter were in the gift shop. He was telling his daughter all about the ghosts and hauntings aboard the ship. He said out loud, "Hey, let's have some fun and try to contact one of these so-called ghosts." He went on to say, "If there is a ghost in here, show us a sign." Immediately, a gift plate went flying off the shelf and landed on the carpet. The detective examined the wire plate holder that was holding the gift plate and it seemed to be in fine condition. He came to the conclusion that the gift plate could not have flown off the shelf by itself.

POOL ROOM SPECTER

Two friends decided to go on the formerly-called Shipwreck attraction that the *Queen Mary* had every Halloween season. The two friends were by themselves near the pool area when something strange happened. All of a sudden, the ghost of a very pale girl in her late thirties appeared. She was dressed in 1940s attire and followed the two friends for awhile. One of the friends relayed that the ghost was getting closer to them and they started to run. The specter managed to keep up with them. The other people that were there did not see anything.

GHOST POKER

Kelly and her family visited the *Queen Mary* when she was a young child. She relayed that she has a very good memory and can remember instances even as far back as childhood. Thus, she remembered the ghost sighting that she had on the *Mary* when she was five or six years old.

Kelly mentioned that her and the family were on Main Deck and were looking around at the ship's stern. There was a room with windows on both the port and starboard sides of the ship. Kelly recalled that she remembers looking into this room and seeing the ghosts of men sitting at a table playing cards. The ghostly men were also wearing old fashioned clothing. A few moments later, the apparitions were gone and no where to be found. Was this a residual haunting or psychic impression of World War II servicemen playing cards to pass the time away?

THE MAID WHO DIDN'T ANSWER

A bunch of family members were onboard the *Mary* in 2003. The adults told the children some of the ghostly tales that abound the ship. The family members walked around some of the exhibits after going on a tour. The family recalled feeling very uneasy and strange when looking into the glass-enclosed exhibits. Later on, they decided to travel down to the engine room.

A couple of family members were on their own when they ran into a possible ghost. As they wandered down hallways, they started to get a very unwelcome feeling. The two family members saw the possible apparition of a woman who looked like she belonged to the housekeeping staff. She had a housekeeping cart as well. The two guests asked her for directions and she only pointed. Thus, she did not say a word. When they reached the end of the hallway, this possible ghost had completely disappeared along with her cart.

It is quite possible that the maid and her cart entered one of the staterooms, which explained her disappearance. If not, then this was most likely another paranormal sighting. Commonsense says that the people would have seen or heard this woman enter one of the staterooms. Meanwhile, one of the other guests had her own unique paranormal encounter. It was more or less a personal experience when she felt the presence of an entity.

When the family returned home, they had their pictures of the *Queen Mary* developed. The picture of the spa contained a little ghostly girl dressed in a blue dress and white bonnet. She was not seen at the time the picture was taken. Could this have been Jackie, as she has commonly been seen wearing a blue dress?

DOOR KNOB JINGLES

Vicki stays aboard the *Queen Mary* quite often. In 1979, she stayed in cabin room A111. She was sick that night and her husband was working. She said that on three occasions, someone came to her stateroom door and wiggled the door knob. Vicki said that she got up and looked down the hallway, only to find that no one was present. She even grabbed the door knob as it was being wiggled by some unseen force. No one was present on the other side of the door. This is another frequently reported claim and an occurrence that I have experienced on a few occasions.

A GHOSTLY GENTLEMAN

B.J. and her mother were taking a summer vacation and spent their usual three nights aboard the ship. While they were there, a male apparition dressed in 1940s attire asked them if they wanted help looking around the ship. B.J. and her mom heard a noise and turned away for a brief moment. When they turned back, the ghostly gentleman was nowhere in sight. There was not enough time for the ghost to exit the dining room. When B.J. and her mother were in the pool room, they saw the same male entity again. He came out of a door and turned and looked at them before walking away.

At 2:00 a.m., they saw this specter again. This time, the two guests witnessed him as he was looking over the railing towards the bay. B.J. and her mother then saw this entity at the front of the ship. Time suggested that there was no possibility for a living person to have out-ran the two women to get to the aft portion of the ship. The very last time the two ladies saw this apparition was in a hallway. He was walking down the hall and de-manifested before their eyes. Was this a former captain making his rounds? Or, perhaps, William Eric Stark?

THE MAN IN DOOR #13

A few years ago, Denise and her husband went for a drive and ended up at the *Queen Mary*. At the last minute, they decided to take a tour of the lower decks. Denise remembers when the *Mary* arrived in Long Beach back in 1967.

As Denise and her husband descended to the lower portion of the ship, anxiety encompassed her. The anxious feeling heightened as she approached water-tight door #13. When Denise walked through the doorway, she turned around and noticed the apparition of a man over her shoulder. This ghostly man was in the doorway one second and gone the next. As he disappeared, her anxiety also lessened. Was she tapping into the energy that J.P. felt on that tragic day in 1966?

GHOSTLY PANIC

On August 23, 1999, a husband and wife were staying on the *Queen Mary* for their wedding anniversary. They took a late-night walk around the decks. At around 1:30 a.m. on the following morning, they arrived at the ship's bow. The female guest reported hearing the frantic screams of men. Was this the residual sounds of screaming sailors during the HMS *Curacoa* accident? Sadly, this is a common reported residual occurrence on the ship.

THE BOY AND THE MUSIC

Kelly was staying a few nights aboard the *Mary* because she was attending a conference. Her friends invited her over to the *Queen Mary* Suite one night. Kelly informed her friends that the ship is haunted, so they shut off the lights to see if they could come into contact with a ghost.

One of Kelly's friends was playing harmonica music when one of the guests saw a blue mist in the doorway leading to the main hallway. The blue mist transformed itself into a little boy who seemed to be attracted to the harmonica music. He especially liked it when "Amazing Grace" was played. All of the guests were in awe as the ghost of the little boy floated toward them in the living room portion of the suite. The apparition disappeared as the harmonica music stopped.

A WEDDING GUEST

Debby attended a wedding that was held at the *Queen Mary*. At approximately 1:30 a.m., Debbie went into one of the dance rooms and stood in the room for about ten minutes. She started to feel the presence of an entity. All of a sudden, a translucent-looking man was standing by one of the columns in the room. She described the apparition as wearing a nice tuxedo with a vest. He was also wearing a top hat and a pocket watch chain. He stood by the column for a few seconds. As Debby walked over to the male ghost, he vanished before her eyes.

GHOST IN THE HALLWAY

Cari and her husband stayed aboard the Queen for their honeymoon. The newlyweds checked into their stateroom and unpacked their belongings. As they got ready for a nice dinner, Cari's husband asked her if she touched him. She told him that she did not do so. He was in the bathroom and she was in the bedroom at the time.

After their dinner, the couple did some sight-seeing and retired to their stateroom. As Cari's husband was unlocking the door to their room, she saw a shadow quickly dart by with her peripheral vision. It sounded like it was running down the hallway. Cari mentioned that she looked out in the hallway and did not see a soul. Logically, there were no places for this person to have disappeared to. Needless to say, Cari said that she felt uneasy for the remainder of her stay. I have had this same type of experience many times on the ship.

STARTLED BY A GHOST

In February of 2000, a boyfriend and girlfriend went to the *Queen Mary* for a vacation. The female guest said that once she stepped foot on the ship, she knew that it was haunted. Thus, she did not want to stay overnight, but her boyfriend convinced her to do so.

While the couple was walking through the ship and looking at the different displays, a man walked through the door at a quick pace. The female guest was so startled that she felt her pulse to make sure that her heart was still beating. When she screamed at the man, he did not flinch or blink an eye. He just casually glanced at the couple and said nothing to them as he walked away. A few minutes later, the female guest said that they saw a picture of the ship's first captain and he was identical to the man they had just encountered. This would be Commodore Sir Edgar Britten, who has been seen by many other guests before.

That same night, the female visitor did not sleep well because she kept hearing weird noises. The following morning, the couple both heard a knock emanating from the closet in their stateroom. After looking and finding nothing in the closet, the boyfriend shut the door. About ten seconds later, the knocking continued. No one was outside the room or present in the hallway. Furthermore, the front desk said that no guests were booked within eight to ten staterooms away. While staying aboard the liner, I have heard knocks on my stateroom doors many times.

ENGINE ROOM GHOST

On March 28, 2001, a husband and wife decided to skip Disneyland and head for the *Queen Mary*. They arrived at the ship at 10:00 a.m. and started touring the engine room exhibits. The male guest told his wife that he felt someone tugging on his arm to make him turn around. He thought another tourist touched his arm; however, the tourist behind him was several feet away. Furthermore, the tourist that was several feet away also had both hands in his pockets.

The couple also relayed that they felt as if several people were walking up and down the stairway by the nursery. They were the only people on the stairway when they felt the presences of other energies. This particular location aboard the ship seems to be consistently active with paranormal activity.

A CURIOUS LITTLE GIRL

On April 6, 2000, Kristina decided to stay on the Queen for some fun. When Kristina and her friend were walking down the Promenade corridor, she felt someone tug on her arm. She looked back and saw a little girl who had a sad expression on her face. As Kristina motioned for her friend to look at the little girl, the child had completely vanished. Kristina's friend did not believe that she had seen the girl. Around 5:30 p.m. that night, Kristina went to her cabin to get some money for dinner when she felt that same familiar tug on her arm. She looked and saw the same small child tugging at her arm and motioning for her to follow. Kristina followed the ghostly child for about three minutes. Did this apparition think that Kristina was her mom?

THE GHOSTLY HAND

About ten years ago, John was in the engine room when he felt a hand gently but firmly grasp onto his left shoulder. The hand slowly pushed John to his left. He let the unseen force turn him until it appeared to go away. As John looked straight ahead, he saw what appeared to be the infamous door #13. If this was the ethereal presence of J.P., what was he trying to communicate to John?

CLOSE ENCOUNTERS WITH GHOSTS

In May of 2001, Angela and her family visited the *Queen Mary*. During their first hour on the ship, Angela felt someone tug at her boot while they were walking down A Deck. Later, near the Promenade Deck, Angela's husband said that he saw a little girl in a white dress. They followed the small child, but she completely disappeared. Around midnight, they returned to stateroom A119 and got settled in for the night. As Angela started to doze, she was awakened by someone tapping on her shoulder. All other family members were fast asleep. Then, Angela felt some unseen force shake her. This occurred for the rest of the night and made sleep quite elusive for Angela. She was kept awake until approximately 7:30 a.m.

A DISEMBODIED GIGGLE

David and Lisa visited the *Queen Mary* in summer of 1991. It was late in the day, so they took an independent guided tour of the ship. As they exited the engine room, they came through a door which led to a small hallway and stairway. David and Lisa reported that they felt very chilly. Then, they both heard the voice of a small child who was laughing and giggling. The child's laughter was so loud that it seemed to be emanating from the same area that David and Lisa were in. The ghostly child's laughter went on for about five minutes. There was absolutely no place for a living child to hide or disappear to. Who was making the laughing sounds? Was this the famed Jackie attempting to make contact with David and Lisa?

THE PINK LADY AND PLAYFUL CHILDREN

In 1999, Trudy and her family were onboard the ship in the Piccadilly shopping area on Promenade Deck. When Trudy entered the small restroom, she came into contact with the smell of perfume. She could hear someone else in the bathroom washing their hands. As Trudy exited the bathroom stall, she noticed that there was a lady in a pink 1920s dress with a hat. The lady was fixing her lip stick. When Trudy came out of the stall, the lady turned, looked at her, and smiled right before she vanished. More interesting was the fact that Trudy's daughter did not see the lady in pink enter or exit the bathroom.

Trudy and her family stayed overnight and saw children playing around midnight. The children disappeared before their very eyes. Trudy mentioned that her husband's mother was adopted in Scotland and brought over via the *Queen Mary* when she was a baby. Thus, these particular ghostly encounters were sentimental.

THE WOMAN IN THE ISOLATION WARD

A teenage girl and her mother frequent the *Queen Mary* quite often. This girl's ghostly sighting took place in the ship's Isolation Ward. The former Isolation Ward was used to treat the sick and injured during the ship's sailing days. Once the teenage girl entered the ward, cold encompassed her body, and she got the chills. Apparently, her mom was unaffected by this encounter. Anyway, there was no vent or open window that could account for the cold feeling.

She glanced over at one of the bunks and noticed a pale young woman who was wearing a purple blouse. Apparently, the sickly woman looked up at the visitor and choked and vomited. Much to the visitor's surprise, the young ghostly woman faded into nothing. Many people have reported having paranormal experiences in the ward. Additionally, many people, including myself, have had headaches and nausea in the location as well.

HOT AND COLD

Jason had the opportunity to visit the ship with the SGVPR paranormal research group. The group felt strong presences in many of the ship's hot spots. However, the weirdest experience took place in one of the boiler rooms. While on the walkway, Jason said that he had the strangest sensation come over him. Part of Jason's body was cold while the other half was hot. The paranormal group scanned Jason's body with thermal scanners, flashing cameras, and video cameras. Additionally, Jason's camera battery had failed completely even though it had been freshly charged. This hot-and-cold sensation lasted for a few minutes. Extreme temperature differences are common when something paranormal is occurring. Normally, most people feel a drop in temperature as opposed to a rise in temperature. This is what makes Jason's experience quite unique.

A GHOSTLY CONVERSATION

California State Fullerton and Cal. Poly Pomona had their dance aboard the *Queen Mary* in 1998. One of the group members had to use the restroom. As the member was in the bathroom, she had a conversation with a female in uniform. After two or three minutes, the woman had left. As the group member's friend asked her what she was talking about in the bathroom, she said that she was having a conversation with a lady that had just entered the bathroom. The visitor's friend said that she did not hear anybody else talking, nor did she see anyone else besides her friend in the restroom. Then who was this mysterious female specter in the bathroom?

THE TRANSLUCENT SPECTER

Emily was visiting California and decided to see the *Queen Mary* with her father. As they were checking out the former Isolation Ward, Emily glanced over at one of the bunk beds. To her amazement, she saw a translucent figure rubbing his neck while he sat on the bed. According to Emily, the apparition was present for quite some time. Emily looked at her father and then looked back at the bed, where the figure had disappeared.

FOOTSTEPS IN THE ENGINE ROOM

About eight or nine years ago, some friends made a trip to the *Queen Mary*. They made their way down to the engine room. Once they arrived in the engine room, it was hard for them to see anything since all the lighting had been completely shut off. After walking around, the friends looked for a stairway that would lead them out of the location. Then, all of a sudden, one of the friends noticed a light at the end of the hallway. A door suddenly opened and footsteps were heard coming down the hallway. The friends thought that it was one of the ship's employees walking down the hallway, so they crunched together and attempted to hide. The footsteps were accompanied by talking. The friends heard two people talking about the weather and the calm condition of the water. This was strange since the ship was docked. The footsteps and voices continued to come closer, and then all of a sudden, they completely stopped. The friends looked around to see if they saw anyone with them. No one else was present besides the group of friends. This was most likely a residual phenomenon from the ship's sailing days. In fact, residual spiritual conversations are reported frequently in the after engine room's shaft alley areas.

THE WOMAN WITH CURLY HAIR

In July of 2001, three friends toured the *Queen Mary.* At around 10:30 p.m. on B Deck, the friends saw the figure of a gauntly woman with curly hair. The woman was wearing white. As she turned to look at the friends, she vanished. This sighting occurred in one of the ship's banquet and meeting rooms. Needless to say, the friends left the ship and did not return that night.

There may be more than one lady-in-white aboard the ship. We know that Tony Ashlin, one of the liner's Paranormal Shipwalk tour guides, has reported sightings of a lady-in-white exiting one of the B Deck staterooms while proceeding to travel aft toward the Isolation Ward area. Thus, I am wondering if this particular curly haired apparition is the same one that Tony has documented.

THE SOLDIER AND THE BOY

Shawn was on the *Queen Mary* one night when he heard footsteps. The footsteps grew louder until he saw the figure of a man in a sailor suit. As Shawn went to the engine room, the lights turned off and he heard more footsteps. According to Shawn, the fire lit up in one of the furnaces. Then, he relayed that he saw a little boy who appeared to be floating. The boy said, "Follow me; it is alright; stay another night." The boy then disappeared.

MYSTERIOUS SOUNDS

J.W. was onboard the ship in January of 1999. He and his wife were occupying one of the suites adjacent to the famed Winston Churchill suite. J.W. was awakened in the middle of the night by faint engine sounds. He also noted that he felt as though the ship was vibrating. This is extremely odd since the ship is docked and its engines have been removed. At first, J.W. thought that these sounds were coming from another ship in the harbor. He got out of bed and looked out the porthole only to find that there was not another ship in sight. It is doubtful that housekeeping does their vacuuming at 3:00 a.m. Additionally, J.W. is familiar with the sounds of ship's engines. This could be another example of residual activity from the vessel's sailing days.

CHILLY HANDS

A husband and wife were staying overnight in one of the first class suites aboard the ship. On their second night's stay, they went to bed quite late. The wife had a difficult time falling asleep, and at one point, she felt cold hands place themselves around her. As she turned toward her husband to say that his hands were chilly, she noticed that he was facing in the opposite direction! Her husband's hands were not even close by. The wife then turned the lights on due to being startled. The couple woke up in the morning to find that the lights had been turned off by themselves. Room lighting turning on and off of its own volition is another commonly reported phenomenon in the hotel decks.

THE BLONDE BOY

A former employee of the ship recalls her experience that she had back in 1998. One day, she noticed a little boy with blonde hair wearing old-fashioned clothes. He was standing there in a ray of light near the Capstan's Room. The ghostly child appeared to be about three to four years of age. The little boy gazed down towards the bow of the ship. The boy disappeared quite fast and the former employee could not arrive at a logical explanation for the child's disappearance. Other people have encountered a small boy with blonde hair as well.

A SICK FEELING

Cindy was visiting her parents back in 1977. Cindy and her parents decided to tour the *Queen Mary*. The propeller room was the spot where Cindy began to feel ill. She broke out in a cold sweat and became nauseous. These are common feelings when a spirit is trying to use a living person's energy. Cindy started to feel hopeless and sensed an impending doom. Once she exited that area, she began to feel normal again. It is most likely that a spirit was using her energy to try and manifest. Or perhaps she was tapping into the fear that the *Curacoan* sailors felt as they experienced their final moments.

THE MAN IN UNIFORM

A visitor to the ship described the ghostly man that she encountered while she was there. She described him as having a moustache and a white uniform. He also had a belt around his waist and red adjusted pants. The visitor also recalled that another ghostly man in the same uniform was present. They were just standing right in the center of one of the staircases.

THE DANCING LADY

On one Christmas Eve, a family made a trip to the *Queen Mary*. When they were down on B Deck, one of the guests reported that she saw a lady. The children also saw the lady as they asked their mom if she saw the woman as well. When they got a closer look of the woman, she had brown hair and was wearing a long dress with white long sleeves. She appeared to be dancing. As they approached the end of the corridor, the female apparition seemed to have de-manifested.

THE CIGAR SMELL AND THE SOLDIER

A married couple spent their 30th wedding anniversary aboard the *Queen Mary.* They stayed on B Deck. They reported that exactly at midnight, their television shut off of its own volition. Then, the bathroom water turned on by itself. The smell of cigar smoke permeated the room, even though no one was smoking a cigar.

On another visit to the *Mary*, the couple had their six-month-old grandson with them. His grandpa told the baby to wave at the nice soldier who was standing in front of them and smiling at the grandson. Apparently, the grandfather and grandson appeared to be the only ones who could see the apparition.

ISOLATION WARD CHILLS

A young girl and her family entered the former Isolation Ward on the ship when they had an unexpected occurrence with the otherworldly. The girl relayed that she sat in one of the former beds when she started to get cold chills. Then she described movement of the bed as if someone else was sitting down on it. The girl was the only person (living, that is) on the bed, and when she looked around she saw the appearance of a disembodied face in the looking glass. She also reported that she saw what appeared like ghostly hands touching the glass from the inside. There were no reflections in the glass that could have explained the girl's encounter.

THE DINING SALON BEAUTY

A woman from Portland, Maine, discussed her odd experience while sitting in one of the *Mary's* dining facilities. As the visitor was sipping tea and munching on toast, she noticed a beautiful red-haired lady across the room. The Maine resident mentioned that the female specter was transparent and wore a long red silk gown complete with a large silk ribbon in the back. She also relayed that the pretty ghost was looking right at her with a grin on her face.

BANGING NOISES IN THE ENGINE ROOM

A few close friends were down in the engine room one night when they encountered something quite odd. They were in the engine area alone when they heard banging noises against all of the antiquated engines. They traced after the noises only to find no one but themselves. When they came to a small door with a lock on it, they noticed that it was open. They also claimed that they felt icy cold air in the vicinity of the door. There were no drafts that could explain the cold spot.

THE GHOSTLY SOLDIER

On the night of December 26, 2003, a couple went to the *Queen Mary* for dinner and the acclaimed Peter James ghost tour. Prior to the start of the tour, the couple walked down to B Deck where they noticed a young man at the end of the hallway. He was dressed in an army green service uniform with short sleeves. The specter just stood there like a statue before he turned and walked straight into the wall. The male visitor ran after the ghost but could find no trace of the young specter dressed in green. Later, they inquired about the attire of the ship's maintenance men only to find out that they wear navy blue pants and jackets with a light blue shirt. Security pointed out that no ship employee wears green clothing.

THE GRAY LADY

A female tourist to the *Queen Mary* discussed her weird experience on the ship many years ago. While on the upper decks of the ship, the tourist relayed that she saw a woman dressed in an antique blue suit dress. This woman had brown hair and just seemed to be walking on the upper decks. The visitor also said that the woman was smiling as she was walking along the decks. Several weeks later, the tourist was watching a television show on the hauntings of the ship. One of the segments of the program showed a woman believed to be the gray lady that travels the ship. The tourist said that she almost fell out of her chair because the face, hair, and smile of the woman she saw walking the decks was identical to the woman featured in the television show.

A BABY'S CRY

A family of three stayed on the Queen for a couple of nights in 2003. After dinner, one of the members took their doggy bag of food to their stateroom when a baby's cry was heard next door. Later on, another member of the family reported hearing the cry as well. Apparently, the room was empty. No one was staying in that particular cabin. I am wondering if this could have been a residual type of haunting from the days the ship carried war brides and children.

A STATEROOM SCARE

A woman had a strange experience in her stateroom one night. She had her four-year-old son with her, so she decided to turn in for the night. At midnight, the mom was awakened by the running water from the faucet in the bathroom. Three hours later, she was awakened again to see that her blankets were being pulled off of her. It would stop and then start up again. At first, she thought it was her son, but it wasn't. The woman said that she was so scared that she grabbed her son and spent the night in her friend's stateroom.

INVISIBLE FOOTSTEPS

Some people were on a tour one night at the *Queen Mary*. When they were in the engine room, they started to hear footsteps coming towards them. However, there was no visible body that accompanied the footsteps. The sounds passed the tourists and continued to approach the other end of the walkway through door #13. What is interesting about this story is that many, many people have experienced this same phenomenon.

TYPES OF HAUNTINGS ON THE RMS QUEEN MARY

The RMS *Queen Mary* exhibits different types of ghosts, spirits, and hauntings. The two types of prolific paranormal activities aboard the ship are considered *residual* and *intelligent* hauntings. A residual haunt can be described as a psychic imprint that is left on the environment, which replays itself in the same location under specific time intervals. It has no intelligence and does not communicate with the living; rather, it is more or less a place memory. A good example of this type of haunting is hearing a phantom cannon or seeing a cluster of Civil War soldiers in Gettysburg. Much of the *Mary's* paranormal activity can fall under this specific category. Two examples of residual activity on the ship are the screams of dying men in the boiler rooms from the HMS *Curacoa* accident and a cluster of ghostly soldiers from World War II marching down the hallways.

Another type of abundant paranormal activity aboard the *Queen Mary* is what paranormal researchers refer to as an intelligent haunting. This is when a ghost or spirit interacts with the living, thus showcasing intelligence, emotion, and thought processes. This can be through any one of our five senses or a combination of them. An example of an intelligent haunting aboard the ship is how the famed spirit Jackie makes a consistent effort to communicate with tourists. Another example can consist of an entity banging on metal in an effort to respond to a proposed question. There are several other types of intelligent interactions between the living and the deceased on the ship.

Another type of haunting that is sometimes witnessed aboard the *Mary* is that of a Poltergeist or psychokinetic (PK) activity. The word Poltergeist is defined as "noisy ghost." In this type of haunting, objects of varying height and weight can be seen moving by themselves. Rarely, people have been tossed across a room by unseen forces. This type of PK energy has been known to start fires, make unexplainable sounds, and cause the levitation of an object or human being.

The common misconception of Poltergeist activity is that it specifically emanates from a ghost or spirit. While this may be true, it is usually the product of an emotionally distraught pubescent teenager. It is theorized that these teenagers are so overcome by strong emotions that they are able to exhibit psychokinetic powers. Thus, the emotions associated with trauma can be brawny enough to move an object.

Many visitors to the *Queen Mary* have reported that their stateroom doors open and close by themselves. Visitors have also witnessed items moving of their own volition. I doubt that there are many pubescent individuals present on the ship at any given time that could suggest the evidence of Poltergeist activity. In my opinion, the strong emotional history of the ship can be enough to produce psychokinetic energy. Also a contributing factor is the fact that the visitors who travel to the liner each day also emit PK energy as well.

Many visitors to the *Queen Mary* have reported incidents which are synonymous to "waking dreams." People who experience these types of dreams are in a hypnagogic state of sleep. This is considered a transitional state from wakefulness to sleep where lucid thoughts and dreams, hallucinations, and sleep paralysis can occur. It is theorized that people may experience paranormal incidents while in this type of slumber.

Some ship guests have relayed that they were dreaming about someone connected to the ship when they awakened to find that same person at the foot of their stateroom bed. Due to the fact that the person is not fully awake, his or her perceptions of what he or she saw become skewed. Thus, it's hard to decipher fact from fiction.

Theoretically, people have been known to have psychic experiences while in the "waking dream" state. Entities seem to utilize peoples' dream states as a way to communicate with them. Many of the intelligent spirits onboard the *Queen Mary* attempt to telepathically communicate with the tourists while they are sleeping. Spirits may find it easier to communicate with the living while they are in a certain sleep state.

No one can really pinpoint what shadow people are or where they come from. They seem to be attracted to places containing mass amounts of human emotion. They are also commonly seen in places that have a lot of paranormal activity. Most paranormal researchers believe that the classic shadow people are in our world strictly to observe the living. It is theorized that these beings emanate from alien species or travel through wormholes. Others suggest that they may be similar to Men in Black. With these beings, people usually report a sense of anxiety or menace associated with encountering them.

In my opinion, I do not think that the majority of shadow-form anomalies seen on the *Mary* correlate to the classic "Shadow People," however. I believe that the ones spotted in various areas of the liner are simply partially manifested apparitions of once living people. Many tourists to the *Queen Mary* report witnessing shadowy figures out of the corner of their eyes. They describe the figures as being either short or tall and resembling human form. The former first- and third-class pool is a hot spot for this type of paranormal energy. Specifically, encounters occur in the upper portside and starboard levels of the pool room.

WHY IS THE QUEEN MARY SUCH A HOT SPOT FOR PARANORMAL ACTIVITY?

SOME GENERAL THEORIES

First of all, I am not entirely sure that any one of us can answer this question. Perhaps the reasons are supposed to remain elusive to us in the mortal realm. The paranormal field has been studied for many, many years. To this day, no one really knows why ghosts and hauntings exist. Thus, there is no black or white logic that can explain why the RMS *Queen Mary* is a Mecca for paranormal activity. What does exist, however, are developed theories as to why the world experiences the paranormal. In my opinion, in order to come closer to understanding what constitutes the supernatural realm, one must study and examine aspects of a person's metaphysical being.

Understanding concepts, such as consciousness and personality, will lay the groundwork for possibly answering the questions that man has had since the dawning of time: "What are ghosts?" "Why does a place become haunted and how does personality survive death?" Learning about the nature of human existence will inevitably help us in comprehending the spiritual world.

In the following pages, I will share a few theories that attempt to explain why places become paranormally active. Furthermore, I will share with you some of the common theories that may expose why the *Queen Mary* is haunted as well as delving into my own personal beliefs as to why the ship is reported to be one of the most supernaturally active places on this planet. It is perfectly okay that some readers may not agree with my personal theories for why the ship is a paranormal hot spot; however, I do want to share my thoughts.

As a paranormal investigator and researcher, I firmly believe that you simply cannot separate a location's history from its ghostly activity. Usually, the paranormal events surrounding a location coincide with its storied past. Thus, historical impressions have been retained in the environment and hold imprints from long ago. For example, battlefields are known to be some of the most paranormally active places on earth. Just imagine all of the emotion and death that are associated with these combat zones. We are not talking about the death of one or two people, but hundreds to thousands of individuals.

Over time, several prevalent theories have been developed that might explain why ghosts and hauntings occur in the living world. Ghosts and spirits seem to be attracted to places that have experienced a lot of emotion, tragedy, and death. Hospitals, convalescent centers, battlefields, prisons, sanatoriums, cemeteries, crime scenes, and churches have all experienced scores of paranormal activity. On the other end of the spectrum, paranormal energy seems to be attracted to places full of happiness and euphoria. This can possibly explain why places such as theatres, taverns, weddings, dance halls, hotels, and theme parks have been known to have ghostly activity.

Another theory as to why places become haunted has to do with ionic energy and the Electromagnetic Field Spectrum. Many paranormal investigators use equipment that can produce ions because it is thought that ionic energy helps a spirit to manifest. It is also thought that a high Electromagnetic Field (EMF) contributes to the growth of ghostly energy. It is theorized that electromagnetic energy can be a foundation for paranormal activity. There are many EMF meters on the market, and they are one of the most invaluable tools that an investigator should have in his arsenal.

Loyd Auerbach is one of the world's most respected and experienced paranormal investigators. In his book, *Ghost Hunting: How to Investigate the Paranormal,* he discusses how Dr. Michael Persinger of Laurentian University in Canada examined how the magnetic field can influence humans and their behavior. Persinger found correlations between psychic experiences and the changing magnetic field. As the magnetic field increased or decreased, people would report psychic experiences. Furthermore, Persinger found that the temporal lobes seemed to be most affected. He developed a helmet for research subjects to wear that would generate magnetic fields, which would then cause them to experience a variety of psychic phenomena.

Antiques and personal belongings may also explain why a location becomes paranormally active. It is theorized that some ghosts are attached to a particular object. Thus, wherever the object is, the ghost or spirit will most likely be right beside it. People who have had an emotional bond to their items want to be near them or make sure that they are kept safe and preserved. This can possibly explain why some homes, museums, antique stores, and libraries become haunted.

Some of the *Mary's* paranormal activity may in fact be partially due to the items used to construct the ship itself. There were over 50 types of wood from all over the world used to decorate the interior parts of the vessel. Thus, could there have been spiritual energy present in the areas and forests that housed the wood? Furthermore, there were certain types of metals used to build the ship. For example, the vessel's propellers contain manganese bronze. The former first- and third-class pool room's ceiling contains quartzite, which is known to retain spiritual energy.

Places that contain portals and vortices are also known to house ghostly energy. Portals and vortices are kind of difficult to explain and understand. Generally speaking, a portal allows spirits to travel from the living world to the spiritual world. Think of it like a bridge connecting two planes of existence. Many paranormal researchers define a vortex as having a lot of spiritual energy. They are also thought to be one of the steps of spiritual manifestation, basically looking like a tubular funnel containing mass amounts of orbs. They resemble the shape of ropes and camera straps. A vortex assists the entity in being continuously charged with energy. In my opinion, the entire *Queen Mary* serves as a portal to the afterlife.

Unfortunately, many entities are earth-bound and are unable to move on for various reasons. For example, a person who experienced a sudden and/or violent death may not know he or she is passed on and continues to perform life's duties. Ghosts or earth-bounds are also stuck in our world simply due to unfinished business or the desire to complete a mortal mission. Furthermore, some entities make the choice to stay on our plane of existence. Then, there are those who can travel to and from our world when they need to help a friend or loved one. There are many aspects that can explain why an entity becomes earth-bound as opposed to finally going home and reaching the level of spirit.

The existence of Ley lines, railroad tracks, and water sources may also explain why places become haunted. It is theorized that Ley lines may contribute to or cause paranormal activity. They are apparent geological and/or cultural alignments of noteworthy locations. In other words, they are considered to be spiritual or mystical configurations of land forms. There seems to be a common thread between railroad tracks and paranormal activity. Water is a natural conductor of electricity. Locations that contain a lot of water have been known to experience paranormal activity. If ghosts and spirits use electrical energy to manifest, then it could explain why places on or near water produce anomalous spiritual energy. Obviously, this is a very prevalent hypothesis concerning the *Queen Mary.*

In terms of the RMS *Queen Mary*, there are several theories, such as some of the ones mentioned in the previous paragraphs, which could explain the continuous existence of its paranormal activity. The *Mary* has endured a rich history, inclusive of varying human emotion, tragedy, and death. With a history as deep-rooted as the *Mary's*, it is no wonder why people report ghostly phenomena on a daily basis.

Scores of people have traveled aboard the *Mary*. Just think of all the folks who have set foot on her decks and how much emotion may be imprinted on the ship itself. Some crew members, passengers, and soldiers succumbed while traveling aboard the *Queen Mary*. Many POWs physically and emotionally suffered and thus tragically passed away while sailing on the ship.

Many psychics have mentioned that a portal exists in the former first- and third-class pool room's dressing area. As mentioned, I believe the whole ship offers a bridge from our mortal world to the afterlife. It is as if the *Mary* has added doses of ghostly activity due to the possibility of housing a portal. Does this contribute to why there is so much paranormal activity aboard the vessel? These are all relevant theories that many people have proposed.

Like I have mentioned in the previous paragraphs, I don't expect all people to agree with some of my thoughts and opinions as to why the *Queen Mary* is haunted. However, I do want to share them with the reader. Personally, I feel that more than one theory can possibly explain the ship's paranormal activity. I think there are many hypotheses that are suggestive of why the *Queen Mary* appears to be a ghost magnet. However, I feel that there are other reasons that can explain the mystery of the ship.

The RMS *Queen Mary* has such a profound spiritual aura around her. It is very difficult to explain in writing, but I will do my best to relay how I feel when I am aboard the *Mary*. I have often pondered if I am one of few people to feel this way regarding the ship; however, after reading about her history, I have learned that others have also felt the same way that I do when present on the *Mary*.

Every time that I have stepped foot on her decks, I instantly feel her spiritual, peaceful, and serene aura. I feel at one with the universe and feel connected to my inner core more than ever when I am on the ship. Basically, I feel "at home" when aboard this majestic liner. All the day-to-day trials and tribulations seem to be forgotten when I am on the ship. I feel so empowered with love, peace, and tranquility when I visit her. This leads me directly to another hypothesis regarding her spirited essence: considering those who have fallen in love with the liner like I have, is it possible that those individuals, whether they were former crew, passengers, World War II servicemen, war brides, or children, have chosen to remain and/or visit the ship in spirit? If this is the case, then think about all of the ethereal souls that exist on the *Mary* at any given moment and on any given day.

I honestly believe that the ship is alive and does communicate with the living. She feels. She thinks. She loves. I have felt this way ever since I first visited the ship back in 2005. The *Queen Mary* is not just an ordinary liner. Since the dawning of her days, she has been extraordinary in her own right. Furthermore, she has a purpose for her continued existence in retirement. It is her quest to continuously share with visitors her unrivaled career on the seas.

I have theorized that some of the paranormal activity is the *Mary's* way of communicating about her history and immortalizing her living legend. I believe that the *Mary* has a way of knowing who is connected to her. Thus, she communicates more to those who have a spiritual tie to her. This could explain why I seem to continuously acquire a lot of paranormal evidence when present on the ship.

Ships do have a soul. As for the RMS *Queen Mary*, her soul is at one with the universe. I sometimes wonder if the *Mary* is a reincarnation of Noah's Ark simply due to my belief that she has been given a protective and spiritual bubble, which has taken care of her since the dawning of her days. The *Mary* has had a rough history and endured many tragedies that very well could have led to her demise. However, she continuously prevailed and miraculously survived every single near-death catastrophe. I do not think that it was due to coincidence that she survived while on the seas. I believe her spiritual protection helped to ensure her survival so she could continue to keep her legend alive in the ensuing years.

In addition to all the feasible explanations for the *Queen Mary's* paranormal events, there's one, that I believe is at the core of her spiritual happenings and it has to do with her being magnetic. It's often a word that I, as well as many others, use to describe the allure of the ship. There's just something special about this ocean liner that draws people to her and it's something she's been doing since her inception. There is a special connection between the vessel's sentience and the intelligence of her resident spirits. The awe-inspiring sense of love and respect people have for the *Queen Mary* amplifies the relations we have with her otherworldly occurrences.

Magnetism in the physical sense, allows objects via an electrical charge to attract or repel each other. Yet, I want to concentrate on the figurative meaning of the word as it pertains to the ethereal activity aboard the ship. The *Queen Mary* is a sentient energy that for years, has been uniting people and attracting those from all over the world. For some reason, she's almost like a giant time capsule that holds all of her memories within for current and future people to experience. In a way, her past, present, and future fuse together, a notion I believe can explain a lot of her spiritual occurrences and general appeal.

Quantum theory of time may help to explain this a bit further. Although controversial, the block universe theory postulates that there is no present and all situations that exist are correlative to each other. The past is a block of the universe at an earlier place while the future is at a later position. Can the past, present, and future signal each other, similar to the firing between nerve synapses in the human body? People have reported ghostly encounters aboard that correlate to certain events occurring on the ship. For example, a rise in sightings of apparitions dressed in fancy period attire during modern retro-style events, such as New Year's parties. Yes, the power of suggestion may be at play here, but there could also be a fusion of the past and present where they each magnetically attract one another.

There may be other elusive explanations for the *Queen Mary's* iconic paranormal phenomena that we mortals are not permitted to understand. Nonetheless, I believe we are given glimpses into the unique intricacies of her interconnection with her resident spirits.

In the beginning pages of this book, I mentioned the following profound words of British psychic Mabel Fortescue-Harrison. I want to mention them again. On the day of the *Mary's* September 1934 launching, Fortescue-Harrison said, "Most of this generation will be gone, including myself, when this event occurs; however, the *Queen Mary*, launched today, will know its greatest fame and popularity when she never sails another mile and never carries another passenger."

"REVERENCE"

Her reflection through her steep presence, is reminiscent of her reverence throughout time, reaching high above, into love she bestows upon the world.

A reminder of the simplicity of love and kindness is automatically expressed through a smile, an embrace, a lending hand, and the words, 'thank you.'

By Norma Strickland

PART 6

RMS Queen Mary Facts and Statistics

Photo courtesy of Joe Bertoldo

FAMOUS PASSENGERS OF THE RMS *QUEEN MARY*

The following is a partial list of some of the famous passengers who traveled on the *Mary*.

Politicians and Statesmen
Winston Churchill
Anthony Eden, British Foreign Minister
General Dwight D. Eisenhower
Walter S. Gifford, American Ambassador in London

Industrialists and Bankers
Henry Ford Jr.
Huntington Hartford
Aristotle S. Onassis
Michael Todd
Jack Wrather

Distinguished Individuals
William Randolph Hearst
Noel Coward
Hedda Hopper
W. Somerset Maugham
J.B. Priestly
Earl Wilson
Frank Lloyd Wright

Sovereigns and Socialites
Duke and Dutchess of Windsor
Baron and Baroness Elie de Rothschild
King Faisal II of Iraq
Sir Cedric Hardwicke and valet
Lady Hardwicke and maid
Master E. C. Hardwicke

Actors, Actresses, and Cinema Personnel

Bob Hope
Charles Boyer
Don Ameche
Bing Crosby
Leslie Howard
Walt Disney
Mr. and Mrs. Fred Astaire and maid
Samuel Goldwyn
Liberace
Harpo Marx
Lionel Barrymore
Norma Shearer
Mr. and Mrs. Charles M. Chaplin and maid
Jimmy Stewart
Johnny Mathis
Loretta Young
Robert Stack
Jack Warner
Elizabeth Taylor
Spencer Tracy
Mr. and Mrs. Gary Cooper
Sophia Tucker
Greta Garbo
Anna May Wong
Oliver Hardy
Eddie Cantor

FILM AND TELEVISION ON THE RMS *QUEEN MARY*

The following is a partial list of some of the film and television productions that have been filmed aboard the ship.

Film

Assault on a Queen
The Aviator
Dangerous Crossing
Final Voyage
Foreign Correspondent
Hidden Agenda
Mame
Pearl Harbor
Dial "M" for Murder
The Poseidon Adventure
She Spies
S.O.S. Titanic
Someone to Watch over Me
The Talented Mr. Ripley
Thirteenth Floor
Trippin

Television

The Ghost Whisperer
Charlie's Angels
CSI Miami
JAG
Maiden Voyage
Murder She Wrote
Night Stalker
Providence
Quantum Leap
Tidal Wave
X-Files

CAPTAINS AND COMMODORES OF THE RMS *QUEEN MARY*

The following is a list of all the captains that served on the RMS *Queen Mary* during her 31 years of service. In total, there were 33 captains. The date in which they assumed command of the ship is also listed underneath their name.

Commodore Sir Edgar T. Britten
December 1, 1935
Captain George Gibbons
January 29, 1936
Commodore Reginald V. Peel
August 4, 1936
Commodore Robert B. Irving
November 11, 1936
Captain John C. Townley
March 30, 1937
Captain Peter A. Murchie
April 19, 1938
Captain Ernest M. Fall
April 9, 1941
Commodore Sir James Bisset
February 23, 1942
Commodore Cyril G. Illingworth
August 10, 1942
Captain Roland Spencer
July 29, 1944
Commodore Chas M. Ford
March 11, 1946
Commodore George E. Cove
December 6, 1946
Commodore Sir C. Ivan Thompson
February 15, 1947

Captain John A. MacDonald
March 6, 1947
Captain John D. Snow
July 4, 1947
Commodore Harry Grattidge
December 31, 1948
Captain Harry Dixon
July 20, 1950
Captain Robert G. Thelwell
August 13, 1951
Captain Donald W. Sorrell
August 19, 1952
Commodore George G. Morris
June 27, 1956
Commodore Chas S. Williams
June 25, 1957
Captain Alexander B. Fasting
September 11, 1957
Captain Andrew MacKellar
August 26, 1958
Commodore John W. Caunce
October 22, 1958
Commodore Donald M. MacLean
June 24, 1959
Captain James Crosbie Dawson
March 30, 1960
Captain Sidney A. Jones
May 25, 1960
Commodore Frederick G. Watts
August 9, 1960
Captain Eric A. Divers
June 19, 1962
Commodore Geoffrey T. Marr
May 7, 1964

Captain John Treasure Jones

September 8, 1965 (Captain John Treasure Jones also commanded the ship on the final 1,001 voyage on October 31, 1967.)

Captain William E. Warwick

September 15, 1965

Captain William J. Law

May 3, 1967

IMPORTANT DATES

Keel Laid: December of 1930

Construction Halted: December of 1931

Construction Resumed: April of 1934

RMS *Queen Mary* Launching: September 26, 1934

Departure from Clydebank: March 24, 1936

Trial Runs: April of 1936

Inaugural Cruise: May 14, 1936

Maiden Voyage: May 27, 1936

Last Pre-War Sailing to New York: August 30, 1939

World War II Service: March of 1940 through September of 1946

War Bride Service: February through September of 1946

First Post-War Sailing: July 31, 1947 (Southampton to New York)

Stabilizer Fittings: 1957 through 1958

Final Voyage (1,001 Voyage): October 31 through December 9 of 1967

Ownership Change: City of Long Beach declares ownership of RMS *Queen Mary* at 10:00 a.m. on Monday, December 31, 1967. (Also reported as December 11, 1967).

Re-fitting Period to Become a Museum and Attraction: January, 1968

Officially Opened as a Museum and Attraction: May 8, 1971

INTERESTING STATISTICS

Gross tonnage: 81,237 (1947); 80,774 (1936)
Length: 1,019.5 ft
Beam: 118 ft
Number of funnels: 3
Number of masts: 2
Number of boilers: 24
Horsepower: 160,000 shp (later became 200,000 shp)
Propulsion: quadruple screw
Regular speed: 29 knots
Highest speed: 32.66 knots
Radar: Two units, ranging up to 40-plus miles
Anchors: Three with a weight of 16 tons
Name Letters: QUEEN MARY on bows
Rivets: 10,000,000
Portholes: 2,000
Public Rooms: 38
Whistles: Three
Watertight Bulkheads: 18
Rudder: 140 tons
Propellers: Four
Engines: Four sets of single-reduction geared turbines with 160,000 shaft horsepower.
Boilers: 24
Light Fittings: 12,000
Clocks: 700
Telephones: 600
Radio Stations: 2
Carpet Range: 10 miles
Fire Hydrants: 378
Swimming Pools: 2 with each of them taking 25 minutes to fill or empty
Oil Usage: 1,020 tons everyday
Normal Speed: 28.5 knots

DISTANCES TRAVELED

More than 3 million nautical miles.
661,771 nautical miles during World War II service.
14,559 nautical miles on final voyage.

PASSENGER CAPACITIES

1936: 776 first class, 784 second class, 579 third class
1946: 711 first class, 707 second class, 577 third class
1960s: 626 first class, 760 second class, 562 third class
World War II service: Up to 16,683 persons
Officers and Crew: 1,285 persons
Lifeboat passenger capacity: 145 persons

The following crew statistics are provided by the book *The Mary: The Inevitable Ship* by Neil Potter and Jack Frost.

DECK DEPARTMENT PERSONNEL (190 PERSONS)

Master: 1
Staff Captain: 1
Navigating Officers: 8
Radio Officers: 12
Carpenter: 1
Assistant Carpenters: 4
Bosun: 1
Bosun's mates: 3
Masters-at-arms: 8
Fire Patrol: 6
Storekeeper: 1
Lamp Trimmer: 1
Quartermasters: 6
Able Seamen: 57

Ordinary Seamen and Boy Seamen: 8
Purser: 1
Staff Purser: 1
Assistant Pursers: 16
Assistant Female Pursers: 4
Baggage Masters: 3
Interpreter: 1
Printers: 5
Musicians: 20
Photographers: 3
Telephonists: 4
Surgeons: 2
Nursing Sisters: 4
Physiotherapist: 1
Dispenser: 1
Hospital Attendants: 3
Officers' Stewards: 3

CATERING DEPARTMENT (829 PERSONS)

Chief Steward: 1
Second Steward: 1
Restaurant Managers: 2
Bedroom Stewards: 81
Waiters and Commis Waiters: 204
Officers' Stewards: 32
Night Stewards and Assistants: 36
Deck and Public Room Stewards: 39
Kitchen Staff: 158

FEMALE STAFF (73 PERSONS)

Stewardesses: 44
Nursery Stewardesses: 3

Bath Attendants: 10
Shop Assistants: 7
Swimming Pool Attendants: 2
Hairdressers: 6
Turkish-bath Attendant: 1

MODERN STATISTICS

City of Long Beach paid $3.45 million for RMS *Queen Mary*.
Number of original hotel staterooms: 346
Current Speed: 0 knots
The ship rises and falls with the tide. The high tide could be as high as 5.5 feet in the Long Beach interior harbor.
Pier E in the Port of Long Beach was the point of arrival for the ship at the end of its final voyage.
Interior wood veneers: 56

IN MEMORIAM: THE CREW MEMBERS OF THE HMS *CURACOA*

ABBOTT, Robert S, Able Seaman, C/JX 235175, MPK
ABBOTT, Ronald B, Ty/Act/Leading Stoker, C/KX 115226, MPK
ADAMS, Walter E, Able Seaman, C/JX 173051, MPK
ADAMSON, Alexander F, Stoker 1c, C/KX 105391, MPK
ADLEN, George B, Officer's Cook 1c, C/LX 21021, MPK
ALDRED, James, Ordinary Seaman, C/JX 351443, MPK
ALEXANDER, Maurice J, Stoker 1c, C/KX 114720, MPK
ALLCROFT, David F, Ordinary Seaman, C/JX 299044, MPK
ALLEN, James E, Ordinary Seaman, P/JX 321862, MPK
ANGER, Harold, Chief Petty Officer Telegraphist, C/J 40064, MPK
APPLEBY, Denis, Ordinary Seaman, C/JX 351445, MPK
ARMITAGE, Thomas D, Ordinary Seaman, C/JX 351446, MPK
ASH, Kenneth R, Leading Cook (O), C/MX 65016, MPK
ASHTON, George W, Ordinary Seaman, C/JX 316156, MPK
ATKINSON, Geoffrey C, Ordinary Seaman, C/JX 351675, MPK

ATKINSON, James J. B., Marine, CH/23783, MPK
ATTOE, Percy J, Able Seaman, C/JX 151752, MPK
AULTON, John, Able Seaman, RNSR, C/SR 59075, MPK
BAILEY, Stanley A, Able Seaman, C/JX 203323, MPK
BAIRSTO, John A, Able Seaman, C/JX 207285, MPK
BAKER, George A, Stoker 2c, C/KX 147052, MPK
BAKER, Gordon J. E., Stoker 1c, C/KX 134339, MPK
BAKER, Ronald J, Ordinary Seaman, C/JX 351184, MPK
BARBER, Sidney F, Marine, CH/X 100195, MPK
BARDEN, Dennis, Ordinary Seaman, C/JX 318882, MPK
BARNET, John T, Ordinary Seaman, C/JX 318875, MPK
BARRETT, Herbert W, Stoker 1c, C/KX 134467, MPK
BARRETT, Kenneth R, Ordinary Seaman, C/JX 351451, killed
BARRETT, Robert A, Able Seaman, RNVR, C/HD/X 40, MPK
BARROW, John H. D., Marine, CH/X 1827, MPK
BATES, Stanley, Ordinary Seaman, C/JX 351450, MPK
BATEY, Thomas W. B., Ty/Leading Cook, C/MX 63356, MPK
BEALE, Albert R, Leading Seaman, RFR, C/J 92989, MPK
BEARD, Rowland D, Engine Room Artificer 4c, C/MX 56948, MPK
BEATON, Peter, Engine Room Artificer 4c, C/MX 76194, killed
BEATTIE, James, Ordinary Seaman, C/JX 251017, MPK
BEEBY, Harold F, Ordinary Seaman, C/JX 317626, killed
BENTON, George, Stoker 1c, C/JX 211125, MPK
BERMAN, William H, Marine (Pens), CH/23494, MPK
BINGHAM, Thomas E, Ordinary Seaman, C/JX 351452, MPK
BISHOP, Philip K, Chief Engine Room Artificer, C/M 35328, MPK
BIXBY, George F, Stoker 1c, C/KX 120562, MPK
BLOTT, Arthur W, Stoker 2c, C/KX 147054, MPK
BLUNT, Lewis F, Ty/Act/Leading Stoker, RFR, C/K 51758, killed
BOARD, Rowland D, Engine Room Artificer 3c, C/MX 56948, MPK
BODGER, Douglas H. J., Lieutenant, RNVR, MPK
BOWLES, Frank D, Ty/Act/Leading Stoker, C/KX 106929, killed
BRAYSHAW, Alfred, Ordinary Seaman, C/JX 351459, MPK
BREAKELL, Edward, Ordinary Seaman, C/JX 351460, MPK
BREWER, Alec A, Ordinary Seaman, P/JX 349726, killed
BRITTAIN, Norman A, Lieutenant, RNVR, MPK
BROAD, William E. R., Stoker 1c, C/KX 109258, MPK

BROCKLESBY, Philip W, Ty/Sub Lieutenant, RNVR, MPK
BRODIE, Oswald, Stoker 1c, C/KX 117162, MPK
BROWN, Kenneth E, Marine, CH/X 3528, MPK
BROWN, Russell H, Ty/Act/Petty Officer, C/J 114792, MPK
BROWN, William R, Ordinary Seaman, RDF, P/JX 350016, MPK
BROWNSETT, Frank, Ordinary Seaman, C/JX 316182, MPK
BRUCKSHAW, Leslie, Ordinary Seaman, C/JX 351464, MPK
BRYAN, Gordon E, Ordinary Seaman, C/JX 316021, MPK
BULL, Harry F, Able Seaman, C/JX 189323, MPK
BULMAN, John R, Engine Room Artificer 4c, C/MX 59009, killed
BUNDAY, John G, Able Seaman, C/JX 189365, MPK
BURTON, Geoffrey D, Ty/Leading Supply Assistant, C/MX 66777, MPK
BURY, John F, Ordnance Artificer 5c, C/MX 56091, MPK
BUTLAND, Edgar J, Chief Engine Room Artificer, C/M 35089, MPK
BUTLER, Albert T, Chief Petty Officer Stoker, C/K 55519, MPK
BUTLER, James H, Able Seaman, C/JX 202896, MPK
CAIN, Reginald N, Able Seaman, C/J 78777, MPK
CALDWELL, Robert, Able Seaman, RDF, P/JX 207935, killed
CALDWELL, Stanley B, Ty/Sub Lieutenant, RNVR, MPK
CALEY, Kenneth, Able Seaman, RNVR, C/HD/X 112, MPK
CANEY, Robert A, Able Seaman, RFR, C/J 107852, MPK
CARD, Francis A. W., Sick Berth Attendant, C/MX 69650, MPK
CAY, Maurice, Surgeon Lieutenant Commander, MPK
CECIL, Thomas, Ty/Petty Officer (Pens), C/J 43783, MPK
CHALLIS, Edward R, Ty/Act/Leading Stoker, C/K 55765, MPK
CHAPPLE, Robert L, Able Seaman, C/JX 238762, MPK
CLARK, Patrick F, Able Seaman, C/JX 150309, killed
COLDRON, Walter R, Stoker Petty Officer, RFR, C/K 65893, MPK
COLE, John S, Act/Captain, RM, MPK
CONEY, Cyril, Supply Chief Petty Officer, C/M 37885, MPK
CONLAN, Peter, Stoker 1c, C/KX 135129, MPK
COPELAND, Cyril, Ordinary Seaman, C/JX 346813, MPK
CORDEN, Frederick J, Able Seaman, C/JX 205102, MPK
CORNELIUS, Robert E, Marine, CH/X 3461, MPK
CORNELL, Percy V, Yeoman of Signals, C/JX 132321, killed
COTTAM, Hubert, Able Seaman, C/JX 237605, MPK
COULSON, Samuel, Able Seaman, C/JX 240712, MPK

COX, Alfred H. S., Ordinary Seaman, RDF, P/JX 358298, MPK
COX, George H. C., Chief Ordinance, C/M 36713, killed
CRAWFORTH, Stanley, Ty/Act/Petty Officer, RNVR, C/HD/X 48, MPK
CREHAN, William L, Corporal, RM, CH/X 1695, MPK
CRICK, Frank D, Marine, CH/X 101769, MPK
CROUCH, James W, Ty/Act/Leading Stoker, C/K 66493, MPK
CUNLIFFE, George P. W., Shipwright 1c, C/MX 47481, MPK
CUTHILL, David A, Supply Assistant, C/MX 81040, MPK
DAVIS, Ernest F, Marine, CH/X 106087, MPK
DAWES, Alfred E, Ty/Paymaster Lieutenant, RNVR, MPK
DEAL, Leonard G, Supply Assistant, C/MX 83008, MPK
DEAN, Harry, Able Seaman, P/JX 304317, killed
DEAN, Robert F, Stoker 2c, C/KX 147065, MPK
DEARSON, Charles G. J., Able Seaman, C/JX 206996, MPK
DELAMAINE, Robert A, Stoker 2c, C/KX 147068, MPK
DOCHERTY, James, Marine, CH/X 100118, MPK
DONOVAN, Jeremiah, Ty/Leading Stoker, C/KX 85041, MPK
DOUGLAS, Robert G, Signalman, C/JX 224194, MPK
DOWNER, Joseph A, Master at Arms, C/M 39869, MPK
DREW, Frederick, Ty/Petty Officer Steward, C/L 14802, MPK
DRIVER, Stanley F, Leading Stoker, C/KX 90453, MPK
DUNNING, Royston, Ordinary Seaman, RDF, P/JX 349472, killed
DURING - ROWE, Walter, Canteen Manager, C/NX 65, MPK
DYASON, Robert G, Stoker 2c, C/KX 137482, MPK
EATON, Donald E, Act/Leading Signalman, C/JX 144843, MPK
EDDY, William, Ty/Act/Leading Stoker, C/KX 84571, MPK
EDMUNDS, Bernard A. J., Ty/Petty Officer Writer, C/MX 60866, MPK
EGAN, Harold, Able Seaman, RNVR, C/HD/X 4, killed
ELLIS, Charles, Ordinary Seaman, C/JX 346793, MPK
ELLIS, Victor H, Marine (Pens), CH/23440, MPK
FERGUSON, Alexander, Engine Room Artificer 4c, C/MX 73700, MPK
FORREST, William A. M., Petty Officer, C/J 105800, MPK
FRASER, Francis R, Marine, CH/23185, MPK
FROST, James E, Act/Gunner, MPK
GARDNER, Daniel H. W., Ty/Lieutenant, RNVR, MPK
GARGET, Sydney, Blacksmith 1c, C/MX 45530, killed
GARNER, Frederick C, Ordinance Artificer 3c, C/MX 58512, MPK

GARWOOD, William F, Ty/Leading Seaman, C/JX 125247, MPK
GAZE, Francis E, Electrical Artificer 5c, C/MX 55239, MPK
GILLINGS, John N, Electrical Artificer 1c, C/MX 46416, MPK
GLACKEN, Patrick O, Ty/Act/Petty Officer, C/SSX 21956, killed
GLANFIELD, William L, Leading Cook, C/M 37713, MPK
GLAZIER, Lewis W, Leading Stoker, C/KX 76281, MPK
GLOVER, Frederick, Able Seaman, C/JX 214721, MPK
GOODBURN, Frederick C, Act/Shipwright 4c, C/MX 53806, MPK
GOWER, Leonard F, Chief Petty Officer Cook, C/MX 46136, killed
GOZZETT, Alfred H, Coder, C/JX 272440, killed
GRANT, Edgar J, Electrical Artificer 5c, C/MX 92737, MPK
GRANT, James L, Warrant Engineer, MPK
GREEN, Albert, Regulating Petty Officer, C/MX 58435, MPK
GREEN, Harold, Able Seaman, C/J 41819, MPK
GROVES, Albert B, Ty/Petty Officer, C/SSX 17166, MPK
HALL, Arthur K, Ordinary Telegraphist, C/JX 216117, MPK
HALL, Charles W, Ordinary Telegraphist, C/JX 233636, MPK
HARDING, George, Ordinary Seaman, C/JX 346927, MPK
HARLING, Peter G, Stoker 2c, C/KX 117869, MPK
HARMAN, John T, Ty/Lieutenant, RNVR, MPK
HART, Ronald J, Able Seaman, C/JX 236917, MPK
HAWKINS, Henry M, Sick Berth Petty Officer, C/M 37726, killed
HAWTHORNE, James, Ty/Act/Leading Seaman, C/JX 151563, MPK
HENDERSON, Robert, Canteen Assistant, NAAFI, MPK
HEPBURN, George, Joiner 4c, C/MX 66747, killed
HEWITT, Sidney H, Petty Officer, C/J 98633, MPK
HEWSON, Arthur, Chief Yeoman of Signals, C/J 72276, MPK
HIDDLESTON, James, Ty/Lieutenant (E), MPK
HIGGIN, John A, Able Seaman, C/JX 169136, MPK
HILL, Albert E. S., Leading Telegraphist, RFR, C/J 49972, MPK
HOLDER, Leslie, Stoker Petty Officer, C/KX 77841, MPK
HOLMAN, Harry A. G., Able Seaman, C/JX 193227, killed
HOLTBY, Phillip L, Able Seaman, RNVR, C/HD/X 7, killed
HOOPER, James T, Able Seaman, RFR, C/SS 11049, killed
HORNER, Kenneth, Signalman, C/SSX 30455, MPK
HOUSTON, William J, Cook (S), C/MX 92438, MPK
HOWARD, Henry, Able Seaman, C/JX 125052, MPK

HOWARD, Roy, Ordinary Seaman, C/JX 346873, MPK
HOWE, Freddy, Able Seaman, C/JX 172658, MPK
HOWLAND, Arthur R, Ty/Paymaster Sub Lieutenant, RNVR, MPK
HOWLETT, Edward H. J., Stoker 1c, C/KX 122494, killed
HULME, Harry, Steward, RNSR, S/SR 8756, MPK
HUNT, William H, Ty/Stoker Petty Officer, RFR, C/K 63830, killed
HUNTER, Harold G, Stoker 1c, C/KX 130132, killed
HUTCHISON, George, Cook (O), C/MX 81820, MPK
HYLTON, Reginald W, Marine, CH/X 1635, MPK
INGALL, William H, Leading Seaman, C/J 81390, MPK
IVESON, Basil T, Able Seaman, RDF, P/JX 197565, MPK
JAMES, Edwin G. H., Engine Room Artificer 4c, C/SMX 59, MPK
JAMFREY, James, Ty/Act/Stoker Petty Officer, C/K 17091, MPK
JANAWAY, James H, Act/Able Seaman, C/JX 257413, MPK
JEEVES, Ernest E, Ordinary Seaman, C/JX 345798, MPK
JEFFERY, Frederick J. W., Chief Petty Officer Stoker, C/K 62836, MPK
JEFFREY, Sidney T, Marine, CH/X 976, MPK
JENNINGS, Edward A, Able Seaman, RNVR, C/HD/X 13, MPK
JOHNSON, Anthony P. C., Lieutenant, MPK
JOHNSON, Stanley, Officer's Cook 1c, C/LX 20344, MPK
JONES, George, Ordinary Seaman, C/JX 346696, MPK
JONES, John W, Ordinary Seaman, C/JX 346693, MPK
JOYCE, Ralph P, Act/Leading Telegraphist, C/JX 178066, MPK
KEEVILL, Raymond C. T., Ordinary Seaman, RDF, P/JX 349472, MPK
KELDAY, Jerry W, Seaman, RNR, C/X 19238, MPK
KIRKLAND, Douglas, Able Seaman, C/JX 197528, MPK
KNIGHT, Robert J, Mechanic 2c, C/KX 78320, MPK
KNOWLES, Albert T, Chief Painter, C/M 37986, MPK
LAKER, William, Stoker 1c, C/KX 130176, MPK
LANE, Stanley C, Ordinary Seaman, RDF, P/JX 321114, MPK
LAZARUS, Leonard, Leading Radio Mechanic, P/MX 89463, MPK
LEASK, Frederick J, Ordinary Seaman, RDF, P/JX 342866, MPK
LLOYD, William J, Able Seaman, C/JX 198531, MPK
LOCKYER, James A, Petty Officer, C/JX 178322, MPK
LOVE, Eric A, Marine, CH/X 2744, MPK
LOVEJOY, James A, Marine, CH/X 106091, MPK
LOW, Walter J. H., Stoker Petty Officer, C/K 64776, MPK

LUGAR, Christopher E, Canteen Assistant, C/NX 2409, MPK
LYALL, Donald F, Ordinary Seaman, C/JX 346839, MPK
MACHIN, Edward H, Ty/Act/Warrant Electrician, MPK
MACIVER, Donald, Able Seaman, C/JX 259465, killed
MACLAREN, Thomas A, Chief Petty Officer, C/J 94539, MPK
MADDISON, George, Leading Stoker, RFR, C/KX 99575, MPK
MALTBY, Francis A, Cook, C/MX 92451, MPK
MANN, Fred S, Marine, CH/X 1559, MPK
MANN, Frederick H, Stoker 2c, C/KX 136787, MPK
MARTIN, John P, Able Seaman, C/SSX 25559, MPK
MASON, Frederick F, Cook, RNSR, C/SR 61338, MPK
MASON, John, Coder, C/JX 229685, MPK
MATTHEWS, Sydney, Able Seaman, C/JX 195691, MPK
MAXWELL, John, Lieutenant, MPK
MCDONALD, Charles, Stoker Petty Officer, C/K 65691, MPK
MCHARDY, Roderick P, Telegraphist, C/JX 232064, MPK
MCLEAN, Thomas, Engine Room Artificer 4c, C/MX 76384, MPK
MEIKLE, John, Ordinary Seaman, C/JX 317857, MPK
MOIR, William, Stoker Petty Officer, C/KX 77485, MPK
MONK, Reginald J, Sick Berth Attendant, C/MX 65419, MPK
MOODY, John J, Ordinary Seaman, C/JX 316927, MPK
MORGAN, Francis J, Ordinance Artificer 4c, C/MX 65549, MPK
MORRIS, James L, Able Seaman, RNVR, C/HD/X 17, MPK
MOSES, John, Stoker 1c, C/KX 129583, MPK
MUMFORD, James F.C., Stoker 2c, C/KX 135643, MPK
MURLEY, Harold G, Ordinary Seaman, C/JX 317477, MPK
MURRAY, William C, Ty/Act/Leading Seaman, C/J 111921, MPK
MYERS, Richard J, Ty/Petty Officer, RFR (Pens), C/J 58522, MPK
NAUGHTON, Douglas J, Ty/Surgeon Lieutenant, RNVR, MPK
NEATHAM, Jack M, Assistant Cook, C/MX 94515, MPK
NICHOLL, George A, Able Seaman, RDF, P/JX 258837, MPK
NICHOLLS, Ronald G, Stoker 2c, C/KX 136492, MPK
O'CONNOR, Terence, Marine, CH/X 103748, MPK
OSBORNE, Stanley R, Ty/Sub Lieutenant, RNVR, MPK
OSELTON, Ronald, Ordinary Seaman, RDF, P/JX 315572, MPK
PARISH, John, Leading Seaman, C/SSX 17875, MPK
PHILBURN, Donald, Stoker 1c, C/KX 127001, MPK

PIMLEY-POPE, Cyril A, Ordinary Seaman, C/JX 299478, MPK
PINKERTON, Joseph E, Stoker 2c, C/KX 110207, MPK
PITT, William, Stoker 2c, C/KX 136391, MPK
PORTER, Arthur H, Stoker 1c, C/KX 127003, MPK
POWLEY, Thomas J, Cook (S), C/MX 65032, MPK
QUEST, Alfred C, Ty/Supply Petty Officer, C/MX 57077, MPK
RAPPALLE, Brian T, Engine Room Artificer 4c, C/MX 60444, MPK
RAWDON, Eric, Leading Radio Mechanic, P/JX 269108, MPK
REASON, William R, Ty/Act/Leading Seaman, C/JX 130616, MPK
REDMAN, George W, Stoker 2c, C/KX 137508, MPK
RHYMES, Albert F, Ordinary Signalman, C/JX 273485, MPK
RICHARDSON, John R, Ordinary Signalman, C/JX 269717, MPK
RICKETTS, John E, Marine, CH/X 100230, MPK
ROACH, William C, Ty/Act/Petty Officer, C/JX 137434, MPK
ROBERTSON, Douglas M, Act/Commander (E), RNR, MPK
ROBINSON, Eric, Ordinary Telegraphist, C/SSX 35111, MPK
ROBINSON, William H, Ordinary Seaman, RDF, P/JX 322082, MPK
SADLER, Charles W, Stoker 1c, C/SKX 1534, MPK
SALISBURY, Harold, Ordinance Artificer 4c, C/MX 69273, MPK
SAUNDERS, Alfred K, Supply Assistant, C/MX 71662, MPK
SAUNDERS, Geoffrey, Marine, CH/X 101595, MPK
SAYWELL, George H, Ty/Act/Petty Officer, RNVR, C/HD/X 73, MPK
SCOTT, Thomas A, Ordinance Artificer 5c, C/MX 56132, MPK
SEARBY, Arthur W, Ty/Act/Leading Stoker, C/KX 115263, MPK
SERVICE, Adam B, Able Seaman, RDF, P/JX 195467, MPK
SETTERFIELD, Charles H, Able Seaman, RFR, C/SS 10302, killed
SHARP, Ronald D, Stoker 2c, C/KX 136498, MPK
SHARP, Thomas W, Able Seaman, RNVR, C/HD/X 28, MPK
SHAW, Alfred W. J., Able Seaman, C/JX 199978, MPK
SHORTEN, George D. J., Petty Officer Steward, C/L 14780, MPK
SHROPSHIRE, Stanley, Able Seaman, C/JX 262395, MPK
SKIDMORE, Stanley, Able Seaman, C/SSX 29628, MPK
SLADE, Frank H, Able Seaman, C/JX 198695, MPK
SMALLSHAW, Jack K, Ordinary Signalman, C/JX 252592, MPK
SMITH, Arthur L, Ordinary Seaman, C/JX 350750, MPK
SMITH, James A. E., Ty/Leading Supply Assistant, C/DX 110, MPK
SMITH, Joe A. B., Able Seaman, C/JX 196735, MPK

SMITH, Reginald C, Able Seaman, C/SSX 27488, MPK
SMITH, Tom G, Stoker 1c, C/KX 114968, MPK
SPEARMAN, Alexander Y, Lieutenant Commander, MPK
SPOONER, Richard C, Chief Petty Officer (Pens), C/J 27956, MPK
STANDING, Thomas G, Electrical Artificer 1c, C/MX 36036, MPK
SUGDEN, Hubert, Able Seaman, RNVR, C/HD/X 30, MPK
SUTTON, Geoffrey C, Paymaster Lieutenant Commander, RNVR, MPK
TADMAN, Charles T, Leading Steward, C/LX 23176, MPK
TAYLOR, Cyril A, Ordinary Seaman, C/JX 301391, MPK
TAYLOR, John, Marine, CH/X 1951, MPK
TAYLOR, Louis R, Ordinary Seaman, C/JX 351857, MPK
THOMAS, Cyril V, Coder, C/JX 229690, MPK
THOMAS, Stanley M, Able Seaman, RNVR, C/TD/X 2048, MPK
THOMPSON, Ernest H. L., Stoker 1c, C/KX 117524, MPK
THOMPSON, Matthew P, Lieutenant, killed
THORBURN, Leslie H, Act/Engine Room Artificer 4c, C/MX 56546, killed
THORPE, John, Able Seaman, RNVR, C/HD/X 32, MPK
TILLEY, Albert J, Schoolmaster, MPK
TODD, James, Marine, CH/X 2161, MPK
TRUNDLE, Eric G, Act/Able Seaman, C/SSX 32498, killed
TURRELL, Frederick, Able Seaman, C/SSX 20725, MPK
TUTTY, Wilson F, Marine, CH/X 3444, MPK
TWOMEY, Jeremiah, Chief Petty Officer Stoker (Pens), C/K 20027, MPK
VAUGHAN, Eric H, Lieutenant, MPK
VERRALL, Reginald R, Stoker 2c, C/KX 117870, MPK
VOYCE, George H, Marine, CH/X 100069, MPK
WALKER, Harry, Marine, PLY/22298, MPK
WALL, Albert E, Chief Petty Officer Stoker (Pens), C/K 32083, killed
WANT, Robert W, Stoker Petty Officer, C/K 60581, MPK
WARD, Frank E, Writer, C/MX 95265, MPK
WARE, James W, Able Seaman, RNSR, C/SR 147, killed
WATKINS, Arthur J, Petty Officer Telegraphist, C/JX 133378, MPK
WATSON, John, Ty/Leading Seaman, C/JX 128078, MPK
WATSON, Thomas W, Able Seaman, C/JX 221743, MPK
WEATHERSTONE, John H, Plumber 1c, RNVR, C/TD/X 263, MPK
WELLS, Edward J, Mechanic 1c, C/KX 82754, MPK
WELLS, Frederick J, Able Seaman, C/JX 215290, killed

WELSH, Sydney E, Ordinary Seaman, C/JX 260086, MPK
WEST, Albert O, Ty/Act/Leading Seaman, C/JX 145528, MPK
WESTRAY, George W, Stoker 2c, C/KX 146846, MPK
WHEELER, James, Stoker 1c, C/SKX 1536, MPK
WHEELER, Vernon J, Stoker 1c, C/KX 115685, MPK
WHITAKER, Edward, Stoker 1c, C/KX 106505, MPK
WHITE, Edward R. G., Sergeant, RM, CH/X 1094, MPK
WHITELAW, George A. N., Ordinary Seaman, RDF, P/JX 248303, killed
WHITTINGSTALL, Wilfred, Able Seaman, RDF, P/JX 263313, MPK
WILKINSON, Roy, Stoker 1c, C/KX 127018, MPK
WILLIAMS, Alun, Stoker 1c, C/KX 116553, MPK
WILLIAMS, Eric, Signalman, C/SSX 31475, MPK
WILLMOTT, Wilfred J, Act/Ordinance Artificer 4c, C/MX 96183, killed
WILSON, Frederick W, Able Seaman, C/SSX 33576, MPK
WINDSOR, William T, Able Seaman, RFR, C/J 87205, MPK
WISE, George L, Marine, CH/X 3485, MPK
WOOD, Alec, Ordinary Seaman, C/JX 318817, MPK
WOOD, Dennis, Able Seaman, RNVR, C/HD/X 81, MPK
WOOD, Geoffrey W, Ordinary Seaman, C/JX 352168, MPK
WOOD, Samuel, Stoker Petty Officer, RFR, C/KX 77407, MPK
WOODCOCK, Sydney, Act/Gunner, MPK
WOODFINE, Frederick W, Able Seaman, RFR, C/J 84725, MPK
WOODS, Denis, Ty/Act/Leading Stoker, C/KX 91462, MPK
WOODS, William, Ordinary Seaman, RNVR, C/HD/X 80, MPK
WOODWARD, Ralph, Coder, C/JX 251538, MPK
WRIGHT, Claude N, Sailmaker, C/JX 134508, MPK
WRIGHT, Jack, Ty/Act/Leading Stoker, C/KX 105626, MPK
YOUNG, Joseph, Able Seaman, C/JX 148378, MPK

IN MEMORIAM: RMS *QUEEN MARY* PASSENGERS, CREW, AND PRISONERS-OF-WAR

The following is a documented list of the individuals who perished while aboard the RMS *Queen Mary*. The names of the individuals are categorized by the year that they passed away.

1934
Malcolm Aitken, Worker
William Eric Stark, Crew

1936
Arthur John Francis Golding, Crew
Commodore Sir Edgar Britten, Crew

1939
Francis Brandt, Passenger
Richard Rolland Metcalf, Passenger

1943
H.M. Holden, Passenger
Faustino Filippini, POW
Captain H.L. Fry, Passenger

1944
P.A.W. Hughes, Crew
John Robert Maloney, Crew

1945
P.H. Ashburn USN, Crew

1946
Leigh Travers Smith, Passenger

1948
J. Robbins, Crew

1949
Herbert Marcus Sichel, Passenger
Charles Summers, Crew
Edwards Clarence Dyason, Crew

1950
Betty Sylvester, Passenger
William Ernest Humphries, Crew
William Henry Barrett, Passenger
Henry Cuthbert Bazett, Passenger
Joseph L. Knapp, Passenger

1951
Alfred John Lee, Crew
Daniel Joseph Martin, Passenger
Kaesar Alexander, Passenger

1952
Harold Healey, Crew
G.T. Joels, Passenger

1953
Morris Shield, Passenger

1954
Mari Anna Ferris, Passenger
George Sydney Wayman, Passenger
Max Sydney Cohn, Passenger
John Owen, Passenger

1955
George Martin, Crew

1956
Kenneth Thompson, Crew
Henry Bonnici, Crew
Walter E. Schott, Passenger
Douglas Alan Thompson, Crew
Mrs. Laurina Sormino, Passenger
Mrs. Olive S. Redfuld, Passenger

1959
Mrs. Frieda Buhlman, Passenger

1963
Steve Radanovich, Passenger
Thomas Chadburd, Crew

1964
Frederick Royal, Crew
Reverend Gerhard Stutzer, Passenger
Mrs. Edith Griffin, Passenger
William Henry Pope, Passenger

1965
A.E. Neville Clark, Passenger
C. McCarthy, Crew

1966
Mrs. C. Nicholson, Passenger
Mrs. Helen Keller, Passenger
Mrs. Honor Teasdale, Passenger
J.P., Crew
Anatole Kundratieff, Passenger
Dr. G.W. McCriley, Passenger
Agnes Lovewell, Passenger

1967
George W. McCreery, Passenger
Captain G.T. Mitchell, Passenger
Arthur Meredith, Passenger
Leonard Horsburgh, Crew

RMS *QUEEN MARY* ROOM CONVERSIONS

The following table lists some of the deck area conversions on A, B, and M Decks. The left column displays the previous room number and the right column displays what the room number changed to.

A is for A Deck
B is for B Deck
M is for M Deck

Previous Room Number	Current Room Number
A 201 and A203	A035
A 200 and A 202	A 036
A 205, 207, 209	A033
A204, 206, 208	A034
A 211, 213, BATH	A031
A215, 217, RESTROOM	A029
A214, 216	A030
A219, 221, BATH	A027
A218, 220	A028
A223, 225, RESTROOM	A025
A222, 224, RESTROOM	A026
A210, A212, BATH	A032
A14	A024
A15	A023
A16	A022
A17	A021
A18	A020
A19	A019
A20	STORAGE
A21	STORAGE
A22	A018
A23	A017
A24	A016
A25	A015
A26	STORAGE
A27	STORAGE
A28	A014
A29	A013

A30	A012
A31	A011
A32	STORAGE
A33	STORAGE
A34	A010
A35	A009
A36	A006
A37	A007
A38	A008
A39	A007
A40	A004
A41	A003
A42	STORAGE
A43	PHONES
A45	A001
STAFF	A002
A48, 50	A102
A49, 51	A101
A52	A104
A53	A103
A54	A106
A55	A105
A56	A108
A57	A107
A58	A110
A59	A109
A60	A112
A61	A111
A62	A114
A63	A113
A64	A116
A65	A115
A66	A116
A67	A115
A68	A118
A69	A117
A70	A118
A71	STORAGE
A72, 74	A120
A73,75	A119

A76	A122
A77	A123
A78	A 122
A80	A126
A81	A125
A82	A128
A83	A127
A84	A130
A85	A129
A86	A130
A87	A129
A88	A132
A89	A131
A90	A134
A91	A133
A92	A136
A93	A135
A94	A138
A95	A137
A96	A140
A97	A139
A98	A142
A99	A141
A100	NO LONGER EXISTS
A101	A143
A102	A144
A103	A145
A104	A146
A105	A147
A106	A148
A107	A149
A108	A150
A109	A151
A110	A152
A111	A153
A112	A154
A113	NONE
A114	A156
A115	A155
A116	A158

A117	A157
A118	A160
A119	A159
A120	A162
A121	A161
A122	A166
A123	A165
A124	A164
A125	A163
A126	A166
A127	A165
A128	A168
A129	A167
A130	A168
A131	A167
A132	A170
A133	A169
A134	A170
A135	A169
A136	A202
A137, BATHS	A205
A138	A202
A139	A171
A140	A172
A141	A171
A142	A172
A143	A173
A144	VENDING/EXIT PASSAGE
A145	A173
A146	VENDING/EXIT PASSAGE
A147	A175
A150	A174
A151	A177
A152	A176
A153	A177
A154	A176
A155	A177
A156	A176
A157	A179
A158	A178

A159	A179
A160	A178
A161	A179
A162	A178
A163	A207
A164	A204
A165	A181
A166	A204
A167	A181
A168	A180
A169	A181
A170	A180
A171	A183
A172	A180
A173	A183
A174	A182
A175	A185
A176	A182
A177	A185
A178	A184
A179	A185
A180	A184
A182	A184
B241, RESTROOM	B341
STAFF/MEN'S RESTROOM	CLOSED
B243, 245, 247	B339
B222, 224, 226	B340
B249, 251, BATH	B337
B228, 230, BATH	B338
B253, 255, 257	B335
B232, 234, 236	B336
B259, 261, BATH	B333
B238, 240, BATH	B334
B263, 265,	B331
B242, 244	B332
B246, 248	B330
B267, 269	B329
B16	STORAGE
B17	STORAGE
B18	B328

B19	B327
B20	B326
B21	B325
B22	STORAGE
B23	STORAGE
B24	B324
B25	B323
B26	B322
B27	B321
B28	STORAGE
B29	STORAGE
B30	B320
B31	B319
B32	B318
B33	B317
B34	B318
B35	B317
B36	B314
B37	B313
B38	STORAGE
B39	STORAGE
B40	B312
B41	B311
B42	B310
B43	B309
B44	B304
B45	B303
B46	B308
B47	B307
B48	B306
B49	B305
B50	B302
B51	B301
B52	B302
B53	B301
B54	B402
B55	B401
B56	B402
B57	B401
B58	B406

B59	B405
B60	B404
B61	B403
B62	B408
B63	B407
B64	NONE
B65	B409
B66	B410
B67	B411
B68	B412
B69	B413
B70	B414
B71	B415
B72	B416
B73	B417
B74	B418
B75	B419
B76	B420
B77	B421
B78	B422 + B424
B79	B423
B80	B424
B81	B425
B82	B426
B83	B427
B84	B428
B85	B429
B86	B430
B87	B431
B88	B432
B89	B433
B90	B434
B91	B435
B92	B436
B93	B437
B94	B438
B95	B439
B96	B440
B97	B441
B98	B442

B99	B443
B100	B444
B101	B445
B102	B446
B103	B447
B104	B448
B105	B449
B106	B450
B107	B451
B108	B452
B109	B453
B111	B501
B112	B454
B113	NONE
B114	B456
B115	B453
B116	B458
B117	B457
B118	B460
B119	B459
B121	B461
B157, 159, 161	B473
B163, B165	B475
B169 & ROOM NEAR HULL	B477
B156, 158 160	B472
B164, 162	B474
B168 & ROOM NEAR HULL	B476
B173, 175	B479
B177, 179	B481
B168	B478
DOCTOR & CONSULT ROOM	B480
B181, 183	B485
B187, 185	B487
B189, 191	B489
B193, 195	B491
B199, 201	B493
B205	B495
ROOM NEXT TO B205	PART OF B497
B180, 182	B484
B184, 186	B486

B188, 190	B488
B192, 194	B490
B200, 202	B492
B206	B494
ROOM NEXT TO B206	B496
B132, 153, 155	B502
B135	B505
B149	BATHROOMS
B151, BATHROOMS	B509
BATHROOMS	B504
BARBER SHOP, BEAUTY PARLOR , STORAGE	B513
BATHROOMS	B513
B196	B513
B197	B506
M1	B515
M2	M037
M3	M038
M4	M035
M5	M036
M6	M033
M7	M034
M8	M031
M9	M032
M10	M029
M11	M030
M12	STORAGE
M13	STORAGE
M14	NONE
M15	M028
M16	M027
M17	M026
M18	M025
M19	STORAGE
M20	STORAGE
M21	M024
M22	M023
M23	M022
M24	M021
M25	STORAGE
M26	STORAGE
M27	

M28	M018
M29	M017
M30	M020
M31	M019
M32	M018
M33	M017
M34	M012
M35	OFFICES
M36	M014
M37	M013
M38	STORAGE
M39	STORAGE
M40	M012
M41	M011
M42	M010
M43	M009
M44	M008
RESTROOMS	M007
M45, 47	M006
M46, 50	LUSITANIA ROOM
M48, 52	M005
M49, 53	M006
M51	M002
M54, 56	M001
M55, 57	M003
M59,61,63	M102
M58,60,62	M101
M65,67 69	M103,105
M64,66,68	M104,106
M71,73,75	M107,109
M70,72,74	M108, 110
M77	M111,113
M76	M112,114
M79,81,83	M115 (Churchill Suite)
M78,80,82	M116
M84	M117,119 (Churchill Suite)
M85	M118,120
M86	STORAGE
M87	M201
M88	M202

M89	M203
M90	M204
M91	M205
M92	M206
M93	M207
M94	M208
M95	M209
M96	M210
M97	M211
M98	STORAGE
OFFICES	M121
M99	M122
M100	M212
M101	M123
M102	M124
M103	M125
M104	M126
M105	M127
M106	M128
M107	M129
M108	M130
M109	M131
M110	STORAGE
M111	M133
M112	M132
M113	M135
M114	M134
M115	NONE
M116	M136
M117	M137
M118	M138
M119	M139
M120	M140
M121	M141
M122	M142
M123 + RESTROOM	M143
M124 + RESTROOM	M144
M125	M215
M126	M216
M127	M145

M128	M146
M129	M147
M130	M148
M131	M149
M132	M150
M133	M151
M134	M152
M135	M153
M136	M154
M137	M155
M138	M156
M139	M159
M140	M160
M141	M157
M142	M158
M143	M159
M144	M160
M145	M161
M146	M162
M147	M161
M148	M162
M149	M217
M150	M218
M151	M219
M152	M220
	M219
B120,122,124,126,128,130,134,136, 138,140,142 removed to construct five new cabins.	M220
	B462,464,466,468,470 are the five new cabins.
B123,125,127,129,131,133,137,139, 141,143,145,147 removed to construct five new cabins.	B463,465,467,469,471 are the five new cabins.

AFTERWORD

"What lies behind you and what lies in front of you, pales in comparison to what lies inside of you."

—Ralph Waldo Emerson

A regal presence, the RMS *Queen Mary* floats tall and proud in the waters of Long Beach harbor. A Queen of the Seas, her unrivaled career and unrelenting success has been a model for British shipbuilding. A stately vessel with a deep soul, she ebbs and flows with the tide and continues to tell her story.

At the beginning of this book, I shared with you that there is something deeply spiritual and mystic about the RMS *Queen Mary*. No amount of written language can explain how I feel and what I feel when I am aboard her decks. All I know is that it is something very beautiful, and probably not meant to be completely understood by any logical standpoint.

As a paranormal researcher, I continue to explore the many pieces of the supernatural puzzle. In regards to the paranormal world, there is so much that man does not yet know and perhaps will never know. If you compare the paranormal field to a beach of sand, I feel that we have only touched upon a grain of that sand in terms of understanding it in its entirety. It knows not black or white, but grey.

With that said, I have many questions surrounding the mysticism of the RMS *Queen Mary*. How long has she been haunted? How many of her former passengers and crew experienced ghostly encounters? If so, what was the first year that ghostly events were documented on the ship? Or do the *Mary's* paranormal events coincide with her permanent docking in Long Beach? My list of questions keeps on growing.

My logical mind wants to find a definitive answer as to why paranormal events occur throughout the *Mary's* inner walls. There are many theories that can explain why, which I shared with you in previous pages. History and the paranormal share a deep kinship and for that reason, I do believe that the *Mary's* history on the seas may have helped lay the foundation for her ghostly events.

Perhaps a few or even all of the aforementioned hypotheses can partly explain why the *Mary* is noted as one of the most haunted places on this planet. When I say partly, I mean that if all the common theories for ethereal existence do in fact apply to the *Mary's* phenomena, there would be other elusive aspects surrounding her mysticism that we would not yet be able to explain.

The RMS *Queen Mary* is a ship with a legendary tale. Many people from different walks of life traveled the oceans on the Queen. Thus, the ship is home to various impressions left from her former passengers and crew, some who are still trying to find their way home. Her inherent beauty, nostalgic character, and abounding history will remain in the heart and soul of her visitors forever.

When I'm aboard, I can feel the *Mary's* soul as she silently whispers to me. She continues to tell the legendary story of her days past, a story that is never-ending. If you intently listen, you, too, will hear the *Mary* as she speaks of her living legend.

O, QUEEN OF THE SEAS

I would like to end this book with the profound words that were written by Norma Strickland, my mother and dear friend.

Born to reign upon the seas
she bridges life from land to land
in breadth and depth, her living breath,
O, Queen of the Seas, she graces me.
From war to peace
her memories
engrained within her inner walls,
the sorrows and the childhood glee
bequeath the whispers near and far.
In presence of integrity
from bow to stern and funnels tall,
she stands within enduring grace
and lures through casts of warm embrace,
upon the waters' majesty
O, Queen of the Seas
she graces me

—By Norma Strickland

CREDITS

Photographs
I want to thank *Queen Mary* enthusiast Joe Bertoldo for allowing me to use several of his amazing photographs of the ship for both the interior and cover. Other historical photographs are courtesy of RMS *Queen Mary* archives.

RMS *Queen Mary* Photo Archives
Joe Bertoldo—*Queen Mary* Enthusiast
R. Villa
Tony Ashlin
Rachel Ashleman
Amy Liam McCallum

Interviews and Contributions
I want to thank RMS *Queen Mary's* Will Kayne for supplying documented paranormal encounters from various people who have visited the ship over the years.

I want to thank the following individuals for their contributions to this book:

Cher Garman: "The Chicago Files," thechicagofiles.wordpress.com
Rosemary Ellen Guiley: Visionary Living Inc., www.visionaryliving.com
Erin Potter: Paranormal Housewives
Norma Strickland: *O' Queen of the Seas*
Patrick Wheelock: *Beyond Investigation Magazine*
Patricia V. Davis: *Cooking for Ghosts,* www.TheSecretSpice.com

BIBLIOGRAPHY

"A Farewell Fete for the *Queen Mary*." *New York Times* 11 Aug. 1967: ProQuest Newspapers. Web. 15 Dec. 2009.

Auerbach, Loyd. *Ghost Hunting: How to Investigate the Paranormal.* Oakland, California, 2004.

Birchall, Frederick T. "*Queen Mary* Begins Maiden Trip Today." *New York Times* 27 May 1936: ProQuest Newspapers. Web. 15 Dec. 2009.

Brown, Jerry. *Long Beach Council to Seek Funds for Queen Mary Repairs.* Travel Tourism and Leisure Collection." Gale. San Diego Public Library. 18 Aug. 2009.

Butler, Allen B. *Warrior Queens: The Queen Mary and Queen Elizabeth in World War II.* Mechanicsburg, Pennsylvania, 2002.

"Cheers Greet Liner after Record Trip." *New York Times* 1 Sept. 1936: ProQuest Newspapers. Web. 15 Dec. 2009.

Deckard, Linda. "Food & drink make up big part of *Queen Mary's* Revenue." *Amusement Business. Hospitality, Tourism and Leisure Collection.* Gale. San Diego Public Library. 15 Dec. 2009.

Dullea, Georgia. "War Brides Relive Days on the *Queen Mary*." *New York Times* 15 Apr. 1985: ProQuest Newspapers. Web. 15 Dec. 2009.

Duncan, William K. *RMS Queen Mary: Queen of the Queens.* Anderson, South Carolina, 1969.

Fenster J.M. "An Ocean Liner Berthed and Burnished." *New York Times* 2 Oct. 1994: ProQuest Newspapers. Web. 15 Dec. 2009.

Graham, Maxtone J.A. "The *Mary*, 1,000 Crossings Later." *New York Times* 3 Sept. 1967: ProQuest Newspapers. Web. 15 Dec. 2009.

Gray, Scott Dr. *The Little Book of Facts about the Big Ship: Details about the Life and Times of the RMS Queen Mary.* 3rd ed. 2003-2007.

Greenbaum, Lucy. "1,666 Brides, 688 Children Here on the *Queen Mary*." *New York Times* 11 Feb. 1946: ProQuest Newspapers. Web. 15 Dec. 2009.

Hinkey, Douglas M. *The Art of the RMS Queen Mary.* Long Beach, CA. 1994.

Horne, George. "*Queen Mary* in with 14,526, who Get Raucous Welcome." *New York Times* 21 Jun. 1945: ProQuest Newspapers. Web. 15 Dec. 2009.

Horne, George. "*Queen Mary* Never Saw a Torpedo as she Roamed Seven Seas Alone." *New York Times* 22 Jun. 1945: ProQuest Newspapers. Web. 15 Dec. 2009.

Kemble, John Haskell. *The Cunard White Star North Atlantic Quadruple-Screw Geared-Turbine Express Passenger Steamship.* Shipbuilder Press, 1936.

"Launching of 534 is Due Wednesday." *New York Times* 23 Sept. 1934: ProQuest Newspapers. Web. 15 Dec. 2009.

"Liner Queen Elizabeth to be Launched Sept. 27." *New York Times* 13 Aug. 1938: ProQuest Newspapers. Web. 15 Dec. 2009.

"Liner *Queen Mary* on Final Trial Run." *New York Times* 15 May 1936: ProQuest Newspapers. Web. 15 Dec. 2009.

Maguglin, Robert O. *The Queen Mary: The Official Pictorial History.* Long Beach, CA. 1985.

Malloy, Betsy. *Efforts to Preserve and Restore the Queen Mary.* California Travel.

Miller, William H. and David F. Hutchings. *Transatlantic Liners at War: The Story of the Queens.* New York, 1985.

Montgomery, Paul L. "*Queen Mary* in Rio on Final Journey." *New York Times* 15 Nov. 1967: ProQuest Newspapers. Web. 15 Dec. 2009.

"Name of New Ship Causes Surprise." *New York Times* 27 Dec. 1934: ProQuest Newspapers. Web. 15 Dec. 2009.

"New Offer Made for *Queen Mary*." *New York Times* 26 May 1967: ProQuest Newspapers. Web. 15 Dec. 2009.

"Offers Pour in for *Queen Mary*." *New York Times* 19 Jun. 1967: ProQuest Newspapers. Web. 15 Dec. 2009.

"13 on the *Queen Mary* were Hurt in Storm." *New York Times* 20 Oct. 1936: ProQuest Newspapers. Web. 15 Dec. 2009.

"Pier is Decorated for Ship Welcome." *New York Times* 1 Jun. 1936: ProQuest Newspapers. Web. 15 Dec. 2009.

Potter, Neil and Jack Frost. *The Mary: The Inevitable Ship.* Great Britain, 1961.

"Records Attained on the *Queen Mary*." *New York Times* 15 Dec. 1935: ProQuest Newspapers. Web. 15 Dec. 2009.

"*Queen Mary* and *Mauretania* as Troop Ships Rumored with Arrival of 770 British Seamen." *New York Times* 19 Mar. 1940: ProQuest Newspapers. Web. 15 Dec. 2009.

"*Queen Mary* Ends Service for U.S." *New York Times* 11 May 1946: ProQuest Newspapers. Web. 15 Dec. 2009.

"*Queen Mary* Ends Shake-Down Trip." *New York Times* 27 Jul. 1947: ProQuest Newspapers. Web. 15 Dec. 2009.

"*Queen Mary* Here with Old Glamour." *New York Times* 6 Aug. 1947: ProQuest Newspapers. Web. 15 Dec. 2009.

"*Queen Mary* Marks Two Years at Sea." *New York Times* 26 May 1938: ProQuest Newspapers. Web. 15 Dec. 2009.

"*Queen Mary* Moved to Long Beach Pier." *New York Times* 18 May 1968: ProQuest Newspapers. Web. 15 Dec 2009.

"*Queen Mary's* Pier Ready for Docking." *New York Times* 31 May 1936: ProQuest Newspapers. Web. 15 Dec. 2009.

"*Queen Mary* Sails; Seen Heading South on War Adventure." *New York Times* 22 Mar. 1940: ProQuest Newspapers. Web. 15 Dec. 2009.

"*Queen Mary* Sets Record for the Atlantic Crossing." *New York Times* 27 Jul. 1936: ProQuest Newspapers. Web. 15 Dec. 2009.

"*Queen Mary* will Remain in California." *New York Times* 11 Oct. 1992: ProQuest Newspapers. Web. 15 Dec. 2009.

"*Queen Mary* Wins in Collision Trial." *New York Times* 22 Jan. 1947: ProQuest Newspapers. Web. 15 Dec. 2009.

Richter, Arthur H. "Cunard Line Starts 120th Year of Trans-Atlantic Ship Service." *New York Times* 5 Jul. 1959: ProQuest Newspapers. Web. 15 Dec. 2009.

Shuit, Douglas P. "Plan Would Send *Queen Mary* to Japan." *Los Angeles Times* 29 Jan. 1997.

"Sir Edgar Britten of *Queen Mary* Dies." *New York Times* 29 Oct. 1936: ProQuest Newspapers. Web. 15 Dec. 2009.

Steele, James. *Queen Mary*. Phaidon Press. March, 2001.

"Steel Work Speeded on the *Queen Mary*." *New York Times* 6 Oct.1935: ProQuest Newspapers. Web. 15 Dec. 2009.

"The City of Long Beach Queen Mary Updates: Queen Mary Critical Repair Efforts." *Long Beach Economic Development,* online, April 24, 2023. https://longbeach.gov/economicdevelopment/real-estate-development/queenmary-updates/

The Cunard-White Star Quadruple-Screw Liner Queen Mary. Bonanza Books. New York, 1979.

The Queen Comes Home. Independent Press Telegram. Long Beach, 1967.

"The *Queen Mary* Getting a Museum." *New York Times* 14 Jul. 1968: ProQuest Newspapers. Web. 15 Dec. 2009.

Thomas, David A., and Patrick Holmes. *Queen Mary and the Cruiser: The Curacoa Disaster*. Annapolis, MD: Naval Institute Press, 1997.

"Transport News: Fast Ocean Trip." *New York Times* 30 Aug. 1966: ProQuest Newspapers. Web. 15 Dec. 2009.

"10,000 Visit the Liner." *New York Times* 4 Jun. 1936: ProQuest Newspapers. Web. 15 Dec. 2009.

Winter, C.W.R. *The Queen Mary: Her Early Years Recalled*. New York, 1986.

"Wives of Soldiers to Come to U.S. Free." *New York Times* 30 Dec. 1945: ProQuest Newspapers. Web. 15 Dec. 2009.

Wlodarski, Robert James and Anne Powell Wlodarski. *The Haunted Queen Mary: Long Beach, California.* West Hills, CA. 2000.

"Work Being Pushed on the *Queen Mary.*" *New York Times* 24 Nov. 1935: ProQuest Newspapers. Web. 15 Dec. 2009.

Zoltak, James. "*Queen Mary* Turns Another Profit Year." *Amusement Business.* Gale. San Diego Public Library. 18 Aug. 2009.

About the Author

One of the leading paranormal researchers on the West Coast, Nicole Strickland is the founder and director of the well-respected San Diego Paranormal Research Society (SDPRS). Since 2011, she has co-hosted the "Spirits of the Adobe" tours at the iconic Rancho Buena Vista Adobe. She serves as the California Coordinator for the Ghost Research Society and is also a consultant to various other investigative groups, including the American Spectral Society.

Blending her love of history, paranormal studies, and writing, Nicole has written several books, including *Field Guide to Southern California Hauntings, The Haunted Queen of the Seas: The Living Legend of the RMS Queen Mary, Spirited Queen Mary: Her Haunted Legend, RMS Queen Mary: Voices from Her Voyages, San Diego's Most Haunted: The Historical Legacy and Paranormal Marvels of America's Finest City, Spirits of Rancho Buena Vista Adobe, Max and Kayli: Two Remarkable Felines Forever Imprinted on My Heart* and *The Afterlife Chronicles: Exploring the Connection Between Life, Death, and Beyond*, which hit Amazon's #1 new release in six categories. *Women on the Fringe: Groundbreaking Women in the Paranormal* debuts in March 2024, an anthology co-authored with Marie D. Jones and Denise A. Agnew. Her books about the *Queen Mary* continue to be best sellers. She is also a writer and contributor to *Paranormal Underground Magazine.*

Nicole is known nationally and internationally for her research on the RMS *Queen Mary* in Long Beach, California. In addition to offering several topics related to the supernatural, she gives presentations about the ship at paranormal conferences, events, and libraries. She has presented at some of the best-known conventions, such as the Oregon

Ghost Conference, Port Gamble Ghost Conference, Troy Taylor's Haunted America Conference, Maritime Ghost Conference of San Diego, Preston Castle Benefit Paracon, Ghost Fest IV aboard the *Queen Mary,* Strange Escapes, OC Paracon, among others. Nicole is represented non-exclusively by RK Entertainment.

Nicole has been featured in a myriad of media outlets discussing her work as a paranormal researcher. These include several local San Diego news programs, such as *Good Morning San Diego*, *San Diego Living*, KPBS Evening Edition, Channel 8 Evening Edition, and Fresno's *Paranormal Journeys.*

Nicole has been interviewed for nationally televised programs, including *My Ghost Story: Caught on Camera*, Travel Channel's *Ghost Stories,* and *Famously Afraid.* She has been interviewed on hundreds of radio shows and podcasts and has appeared in many San Diego newspapers and magazines.

From 2020-2023, Nicole co-hosted *Haunted Voices Radio*, one of the longest-running radio programs featuring a plethora of guests from all areas of the supernatural. In 2020, she debuted her radio show *The Afterlife Chronicles*, which was selected by Feedspot as one of the top 25 programs on the afterlife. She is launching *The Life Inspiration Project* in 2024, a live show / podcast featuring people from all walks of life.

In addition, she enjoys cooking, reading, traveling, and spending valuable time with family and friends. She enjoys working with other paranormal researchers as she believes that through diligent research, teamwork, and collaboration we will all better understand the vast field of the unknown.

Nicole holds a B.A. degree with honors from the University of Arizona (go Wildcats!) and an M.S. degree from National University. She plans to pursue a PhD in parapsychology in 2024.

AUTHOR CONTACT INFORMATION

San Diego Paranormal Research Society
www.sandiegoparanormalresearch.com

Author Nicole Strickland
www.nicoledstrickland.com

Legacy of the RMS *Queen Mary*
www.facebook.com/queenmarylegacy

Facebook—San Diego Paranormal Research Society
www.facebook.com/SDPRS

Facebook—Nicole Strickland
www.facebook.com/nicoledstrickland

Instagram—authornicolestrickland
Twitter—@sdprsnicole

OTHER BOOKS BY NICOLE STRICKLAND

Field Guide to Southern California Hauntings (2009)

First Edition: The Haunted Queen of the Seas: The Living Legend of the RMS Queen Mary (2010)

Spirited Queen Mary: Her Haunted Legend (2017)

San Diego's Most Haunted: The Historical Legacy and Paranormal Marvels of America's Finest City (Schiffer Publishing)

Spirits of Rancho Buena Vista Adobe (The History Press)

RMS Queen Mary - Voices from Her Voyages

Max and Kayli: Two Remarkable Felines Forever Imprinted on My Heart

The Afterlife Chronicles: Exploring the Connection Between Life, Death, and Beyond

Women on the Fringe: Groundbreaking Women in the Paranormal

DO YOU HAVE A MEMORABLE SHIP EXPERIENCE TO SHARE?

If you'd like to share a memorable experience that you've had aboard the *Queen Mary*, please feel free to contact me. I absolutely enjoy reading about visitors' special times aboard the most legendary liner to ever set sail. This also goes for any paranormal experiences that you've had aboard the ship. My Spirited Queen Mary website also has a blog section and your experience(s) could be featured! When describing your paranormal experience(s) aboard the RMS *Queen Mary*, here are some points to remember:

- Time, date, and location of occurrence.
- What you were doing at the time of your encounter.
- Describe the human senses that were involved when experiencing the activity.
- If you witnessed an apparitional sighting, please describe it in detail.
- If you heard a disembodied voice, please describe it in detail.
- If you had an olfactory experience, please describe it in detail.
- If you were conducting an investigation at the time of your encounter, what equipment were you using? Any audio, photographic, or video evidence?
- Do you feel that you witnessed one of the ship's noted spirits? If so, in what location did you have your experience?

- When you had your paranormal encounter, were you a guest on one of the ship's tours? If so, which one?
- Please feel free to describe any intuitive (aka "psychic") feelings you had while on the ship.
- Did you have any telepathic (mind-to-mind) communication with the ship's ghosts and spirits? If so, with whom?
- Do your experiences match those of other visitors to the ship?
- Do your experiences match those described in this book?
- If you saw an apparition or shadow form, can you draw what it looked like?

Made in the USA
Columbia, SC
16 February 2025